# Microsoft® Proje

# Microsoft® Project

by Cynthia Snyder Dionisio, MBA, PMP

## Microsoft® Project For Dummies®

Published by: **John Wiley & Sons, Inc.,** 111 River Street, Hoboken, NJ 07030-5774, www.wiley.com



Published simultaneously in Canada



For general information on our other products and services, please contact our Customer Care Department within the U.S. at 877-762-2974, outside the U.S. at 317-572-3993, or fax 317-572-4002. For technical support, please visit https://hub.wiley.com/community/support/dummies.

Wiley publishes in a variety of print and electronic formats and by print-on-demand. Some material included with standard print versions of this book may not be included in e-books or in print-on-demand. If this book refers to media such as a CD or DVD that is not included in the version you purchased, you may download this material at http://booksupport.wiley.com. For more information about Wiley products, visit www.wiley.com.

Library of Congress Control Number: 2021952565

ISBN 978-1-119-85862-1 (pbk); ISBN 978-1-119-85863-8 (ebk); ISBN 978-1-119-85864-5 (ebk)

SKY10033470_022522

# Contents at a Glance

# Table of Contents

# Introduction

Project management has evolved from a discipline that began with index cards and yarn to one that now uses sophisticated analysis techniques, projections, reporting, and time and resource tracking. Project management software offers functionality that makes planning and tracking the complex projects we undertake a little more manageable.

## About This Book

Microsoft Project is one of the most popular project management software applications. It offers a tremendous amount of functionality to users. However, as with most software, mastering it can seem like a daunting process.

It helps to understand how Project's features relate to what you do every day as a project manager. In *Microsoft Project For Dummies*, my goal is to help you explore Project Professional (an on-premises or desktop version) and Project Online (a cloud-based version). This book provides information on relevant project management concepts while also offering step-by-step instructions to build and track a Project schedule.

Here are some broad topics that this book explores. You can:

- Start out in Microsoft Project by entering tasks and dependencies and estimating durations
- View your project as a Gantt chart, Task Board, network diagram, and other views
- Work with resources, calendars, and costs
- Negotiate constraints, fine-tune the schedule, and set a baseline
- Gather data, analyze progress, take corrective actions, and report project status
- Set up and track a Sprints Project with backlogs, Task Boards, and specialized reports

Throughout this book, I offer advice on how to make all these features and procedures mesh with what you already know as a project manager.

Finally, Microsoft Office runs on Windows 10 and 11 (as of this printing), and not Windows 7 or 8.1, so you'll need to have Windows 10 or 11 in order to run Microsoft Project. This book is written as if you're using the "on-premises" or "desktop client " version of Microsoft Project. If you are using a cloud-based solution, this book assumes the "Project Plan 3" subscription plan. For more information on features for various online subscriptions, you can visit `www.microsoft.com/en-us/microsoft-365/project/compare-microsoft-project-management-software`.

# What's Not in This Book

Microsoft has a lightweight project application called Project for the Web. Project for the Web is appropriate for smaller projects or for people who aren't project managers but who manage projects as part of their job. It is not sufficient for large or complex projects. It is built on the Microsoft Power Platform rather than SharePoint, so the features, functions, and interface are very different. This book does not cover any information about Project for the Web. If you want more information, you find it here: `https://support.microsoft.com/en-us/project`.

Because this book assumes desktop client software, I don't spend time on anything associated with enterprise versions or server features. The good news is, regardless of whether you're running the desktop client or subscription version of the software, this book is packed full of useful information for getting to know Microsoft Project.

# Foolish Assumptions

I've made some assumptions about you, gentle reader. I figure that you're computer-literate and that you know how to use the mouse, the keyboard, and the Ribbon. I assume that you know how to use most common Windows functions (such as the Clipboard), as well as many basic software functions, such as selecting text and dragging and dropping items with the mouse.

I also assume that you have experience in managing projects. Whether you manage very large projects that are several years long or you have been a team lead on a project, the information in this book is more accessible if you have a background in project management.

I do *not* assume that you've used Project or any other project management software. If you're new to Project, you'll find what you need to get up to speed, including information on how Project works, finding your way around the Project interface, and building your first Project schedule. If you've used an earlier version of Project, you'll find out about the current version of Microsoft Project and the features it provides.

## Icons Used in This Book

One picture is worth . . . well, you know. That's why *For Dummies* books use icons to provide visual clues to what's going on. Essentially, icons call your attention to bits of special information that may well make your life easier. The following icons are used in this book.

The Remember icon signals either a pertinent fact that relates to what you're reading at the time (but is also mentioned elsewhere in the book) or a reiteration of a particularly important piece of information that's, well, worth repeating.

Tips are the advice columns of computer books: They offer sage advice, a bit more information about topics under discussion that may be of interest, or ways to do things a bit more efficiently.

Warning icons spell trouble with a capital *T*: When you see a warning, read it. If you're not careful, you might do something at this point that could cause disaster.

## Beyond the Book

In addition to what you're reading right now, this product also comes with a free access-anywhere Cheat Sheet that includes tips on creating your project schedule, shortcut keys, and helpful websites to hone your expertise. To get this Cheat Sheet, simply go to `www.dummies.com` and type **Microsoft Project For Dummies Cheat Sheet** in the Search box.

## Where to Go from Here

Take what you've learned in the project management school of hard knocks and jump into the world of Microsoft Project. When you do, you'll be rewarded with a wealth of tools and information that help you manage your projects much more efficiently.

Your first step might be to read the table of contents and find the sections of this book that you need — or simply turn to Chapter 1 and start reading.

# 1 Getting Started with Project

IN THIS PART . . .

Get familiar with the Ribbon and its tabs in Project.

See how to open a new project, enter tasks, and organize a project.

Become skilled in working with summary tasks, milestones, and notes.

Identify and arrange task dependencies to create a network diagram.

Develop duration estimates for tasks.

Customize Project views to meet your needs.

IN THIS CHAPTER

- » Describing different approaches to managing a project
- » Comparing the project manager and Scrum master roles
- » Benefitting from Project
- » Exploring the software interface
- » Finding help in Project

# Chapter 1 Project Management, MS Project, and You

Welcome to the world of computerized project management with Microsoft Project. If you've never used project management software, you're entering a brave new world.

Everything you used to do with handwritten to-do lists, sticky notes, word processors, and spreadsheets magically comes together in Project. However, this transition doesn't come in a moment, and you need a basic understanding of what project management software can do to get you up to speed. If you've used previous versions of Project, the overview in this chapter can refresh your memory and ease you into a few of the new Project features.

Even if you're a seasoned project manager, this chapter provides the foundation for how to work with Project.

## Project Management Evolution

The profession of project management has evolved significantly in the past 20 years. As a profession project management is more respected and more in demand than ever. Organizations depend on project managers to drive change and deliver value. There is widespread recognition that project management skills aren't just for professional full-time project managers; they can be used by anyone who manages projects as part of their job, even if they aren't in a project management role.

With the rapid growth of technology and technology-driven projects, the way we manage projects has evolved. When Microsoft Project was first released in 1984, projects were plan-driven, meaning that we tried to plan out everything that would happen, in detail, up front. Then we would execute based on that plan. That approach works when you can define the project scope and requirements up front, such as engineering or construction projects. We call this a *waterfall approach* because the completion of one phase leading to the start of another looks like a waterfall, as shown in Figure 1-1.

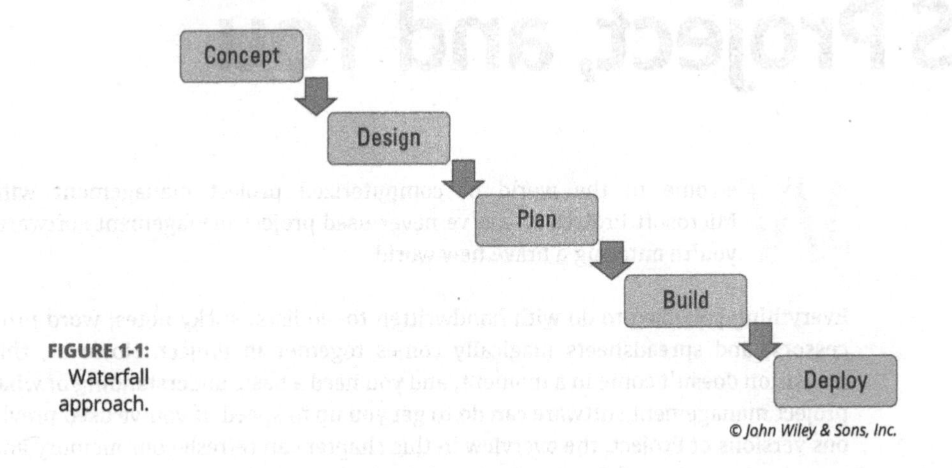


**FIGURE 1-1:** Waterfall approach.

© John Wiley & Sons, Inc.

As the pace of technological growth accelerated, the waterfall approach was no longer effective for technology development projects. By the time you had planned and executed a two-year technology project, the technology had changed, and the end result was already out of date. Therefore, a new approach to managing technology projects evolved.

In early 2001 an approach called *Agile project management* was created. Agile is a mindset that is documented in the Agile Manifesto. It starts with four values:

» Individuals and interactions over processes and tools

» Working software over comprehensive documentation

» Customer collaboration over contract negotiation

» Responding to change over following a plan

Note there is less emphasis on following a plan, and change is expected. This is very different than managing construction projects where following the plan is paramount. There are also 12 principles that are documented in the Manifesto. You can find the Agile Manifesto here: `https://agilemanifesto.org`.

Over time project practitioners have recognized that the type of project you are managing determines the project management practices you should employ. For scope that is stable and can be well-defined up front, a waterfall approach is appropriate. For scope that is not well defined or can evolve based on feedback and market changes, an agile approach (also known as an *adaptive* approach) is appropriate. To make things more interesting, there are many projects with some deliverables that can be well defined and other deliverables that can't — for these projects a hybrid approach that incorporates aspects of waterfall and adaptive approaches is best.

As project management practices have evolved, so has Microsoft Project. This version of Project allows you to set up Gantt charts that are resource loaded to manage well-defined scope and Task Boards to manage evolving scope, all in one glorious piece of software. To keep things simple, I use Parts 1 through 4 to talk about how set up and manage a Waterfall Project and I use Part 5 to show you how to set up and lead an Agile Project. For those of you who work on hybrid projects, you can use both waterfall and Agile functionalities.

Project does use the term Agile. Projects with evolving scope that use adaptive approaches are called Sprints Projects.

# What's in a Name: Projects, Project Management, and Project Managers

In this book, a project is defined as a unique venture undertaken to produce distinct deliverables, products, or outcomes. In the context of a project, a *deliverable* is an individual component or item that meets the requirements of the project, such as a design document or a prototype.

*Project management* is the practice of organizing and managing project variables to meet the project outcomes and mission. Some of the variables we work with are listed in Table 1-1.

**TABLE 1-1** **Project Variables**

| Variable | Description |
| --- | --- |
| Scope | The work needed to produce the deliverables, products, or outcomes for the project. |
| Time | The duration required to complete the project work. |
| Cost | The funds required to complete the project. |
| Resources | The people, equipment, material, supplies, and facilities needed to accomplish the project. |
| Change | *Product* change refers to the features and functions of a product. *Project* change refers to changes in schedule, cost, or resources. |
| Risk | Uncertainty associated with the scope, time, cost, resources, stakeholders, or environment that can threaten the completion of any aspect of the project. |
| Stakeholder | A person who can affect, or who is affected by, the project, either positively or negatively. |
| Environment | The location, culture, or organization in which the project occurs. |

## Project managers and Scrum masters

Projects that use waterfall approaches generally have a project manager who creates the master plan for a project and ensures that it is implemented successfully. A project manager uses technical, business, and leadership skills to manage the completion of tasks and keep the schedule on track. Project teams can be small, mid-size, or very large.

TIP

A truly professional project manager may have a degree in project management or a professional certification. For example, if you see the initials *PMP* beside a name, that person has been certified as a Project Management Professional by the Project Management Institute, the leading global organization establishing project management standards and credentials.

A Scrum master is usually found on projects that use Agile approaches. A Scrum master assists the team in following Agile processes and delivering the work. Teams are optimized at 6 to 10 members.

We look at the roles of project managers and Scrum masters in the following sections.

## The role of the project manager

The project manager is the person who ensures that aspects of the project are integrated and assumes hands-on responsibility for successes as well as failures.

The project manager manages these essential pieces of a project:

- **Leadership:** Documents the project vision in a charter. Facilitates negotiations, problem-solving, and decision-making activities.
- **Scope:** Defines and organizes all work to be done in order to meet the project objectives and create deliverables.
- **Quality:** Identifies, manages, and controls requirements. Establishes a process for quality management and control and ensures it is sufficient and followed.
- **Schedule:** Develops the schedule, by working with Project, including the tasks, relationships, duration, and timing involved to achieve the project objectives.
- **Resources:** Assigns resources and tracks their activities on the project as well as resolves resource conflicts and build consensus. Working with resources also involves managing physical resources such as materials and equipment.
- **Cost:** Estimates project costs and applies those estimates across the schedule to create a time-phased budget.
- **Stakeholder Engagement:** Notifies appropriate *stakeholders* (everyone who has a legitimate stake in its success) of the project status. Facilitates communication with internal and external stakeholders.
- **Uncertainty:** Establishes a system to identify, analyze, respond to, and track project risks and issues. Guides the team in working with uncertainty, ambiguity, complexity, and volatility.

Managing a project requires overseeing all its variables to ensure that the project goals are accomplished on time, within the limits of the budget, and using the assigned resources while also addressing risks, managing change, and satisfying stakeholders.

## The role of the Scrum master

A Scrum master is a servant leader. Servant leaders are focused on supporting their teams, rather than directing them. They educate and support team members in maintaining alignment with Agile practices.

A Scrum master engages in these activities and behaviors:

- **Leadership:** Practices servant leadership. Motivates the team. Provides coaching and mentoring to team members as needed.
- **Scope:** Works with the product owner to communicate the priorities in the project backlog.
- **Schedule:** Facilitates sprint planning, demonstrations, and daily stand-up (or scrum) meetings.
- **Resources:** Protects the team from outside interference. Removes barriers or impediments so team members can accomplish work.
- **Process:** Helps the team and other stakeholders understand and follow agile processes. Works with the team to improve team dynamics and the processes used to create and deliver value.
- **Risk:** Reduces uncertainty by experimenting with different solutions, building prototypes, and providing demonstrations to relevant stakeholders throughout the project.

TIP

A *product owner* determines the vision and scope for the project, makes decisions about the product, and establishes priorities. Where necessary, they interact with external stakeholders to understand needs and then communicate those needs to the team.

You can see that project managers and Scrum masters both use their skills and knowledge to accomplish the project work. The approach is different with the project manager providing more direct oversight and accountability for the outcomes, whereas the Scrum master operates as a supportive role for team members, empowering and enabling them to accomplish the work.

In a hybrid project you may see a project manager for the overall project with a Scrum master working on software development aspects of the project. You might also see the project manager adapting more of the servant leadership behaviors while still maintaining overall accountability for the project. Ultimately, the best approach is one that works in your environment.

Regardless of the approach, having software to help organize and structure the work makes managing the project and leading the team less daunting. That's where Project can help.

# Introducing Microsoft Project

Microsoft Project is a scheduling tool that helps you organize, manage, and control the variables identified in the preceding section. In this book, I show you how to use Project to organize and manage your work, create realistic schedules, and optimize your use of resources.

Take a moment to look at some of the wonderful ways in which Project can help you organize, manage, and control your project:

- **Use built-in templates to get a head start on your project.** Project *templates* are prebuilt plans for a typical business project, such as commercial construction, an engineering project, a new product rollout, software development, or an office move.
- **Organize your project by phase, deliverable, geography, or any other method.** The outline format allows you to progressively elaborate the information in greater granularity depending on how detailed you want your plan to be.
- **Determine costs by your chosen method.** Examples are time period, resource type, deliverable, or cost type.
- **Organize resources by resource type.** Level your resources to avoid overallocation, or determine the impact on the duration of a task based on a change in resources.
- **Calculate costs and timing based on your input.** You can quickly calculate what-if scenarios to solve resource conflicts, maintain costs within your budget, or meet a deliverable deadline.
- **Use views and reports with the click of a button.** A wealth of information is now available to you — and those you report to. You no longer have to manually build a report on total costs to date to meet a last-minute request from your boss.
- **Set up a Sprints Project.** Run a project with a backlog, a Task Board, sprints, and other adaptive practices.
- **Manage complex algorithms** (that you couldn't even begin to figure out on your own) to complete such tasks as leveling resource assignments to solve resource conflicts, filtering tasks by various criteria, modeling what-if scenarios, and calculating the dollar value of work performed to date.

REMEMBER

No matter how cool the tool, you have to take the time to enter meaningful data. Great software doesn't ensure great outcomes; it only makes them easier to achieve.

## Getting to Know You

The file you create in Project is a *schedule model.* It's a model because it models what you think will happen given what you know at the time. However, for ease of reference, I just refer to it as a schedule. The schedule has a plethora of data about various aspects of your project as well as graphical representations of that information.

REMEMBER

Some people refer to the project schedule as the project plan. In reality, the project plan *contains* the project schedule — plus information such as the budget, work breakdown structure, project life cycle, risk management plan, and many other ingredients necessary to effectively manage a project.

When you first open Project, you see several options for starting a new project, as shown in Figure 1-2.

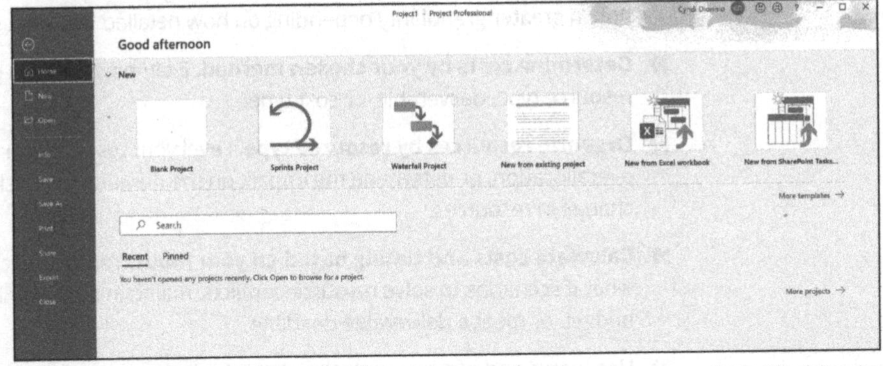


**FIGURE 1-2:** What you see when you open Project.

© John Wiley & Sons, Inc.

You can open a blank project, create a new project from an existing project, or create a new project by importing information from Microsoft Excel or SharePoint. You can also take advantage of premade templates for common project types, such as these examples:

- Residential construction
- Software development
- New product launch
- Merger or acquisition evaluation

If you don't see the template you need, you can click More Templates and, as you can see in Figure 1-3, a whole host of options appears. You can also search for online templates by entering keywords in the Search box.

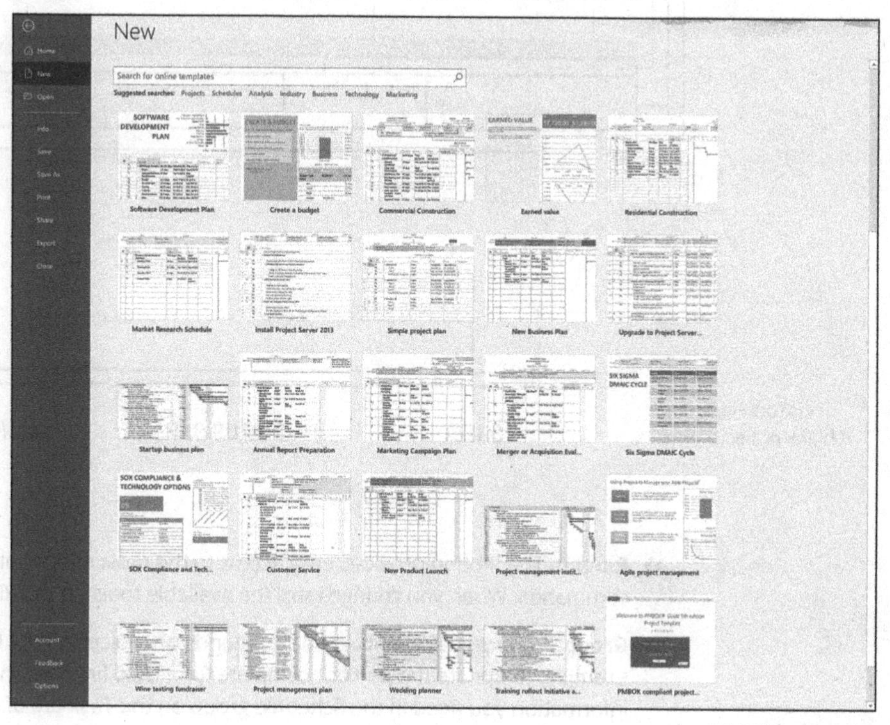


**FIGURE 1-3:** Project templates.

*© John Wiley & Sons, Inc.*

For purposes of this discussion, I assume that you're starting with a new, blank project.

When you open a new project, you see the Quick Access toolbar, a few Ribbon tabs, the Ribbon, the Timeline, a pane with a sheet and a chart, and the status bar, as shown in Figure 1-4.

In Figure 1-4, you see Gantt Chart view. (I discuss views in Chapter 6.) For now, here's an overview of the major elements in Project (refer to Figure 1-4):

- **Quick Access toolbar:** The Quick Access toolbar, above and to the left of the Ribbon, is onscreen at all times and in all views.
- **Ribbon tabs:** The Ribbon tabs organize commands based on a particular type of activity. For example, if you're working with resources, you'll likely find the command or setting you want on the Resource tab.

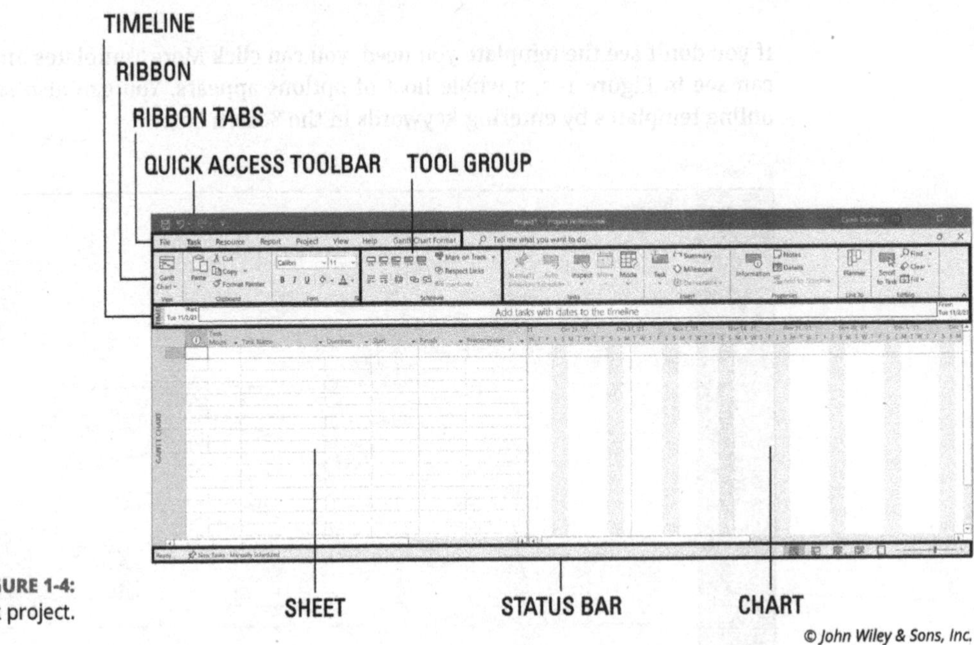


**FIGURE 1-4:** A blank project.



- **Ribbon:** The Ribbon provides easy access to the most commonly used tools and commands. When you change tabs, the available tools on the Ribbon change.
- **Group:** A *group* is a set of related commands or choices on the Ribbon. For example, to update the percent complete for a task, first find the formatting information you need in the Schedule group on the Task tab of the Ribbon.
- **Timeline:** The Timeline provides an overview of the entire project — a graphical view of the project from start to finish. You have the option of showing the Timeline or hiding it.
- **Sheet:** Similar to a spreadsheet, the sheet displays the data in the project. The default fields change depending on the Ribbon tab you're working in. You can customize the columns and fields in the sheet to meet your needs.
- **Chart:** The chart is a graphical depiction of the information on the sheet. Depending on the view or Ribbon tab that's displayed, you might also see a bar chart depicting the duration of a task or a resource histogram showing resource usage.
- **Status bar:** The status bar, at the bottom of the Project window, has information on views and zoom level on the right, and information on how newly entered tasks are scheduled on the left.

## Navigating Ribbon tabs and the Ribbon

Each of the Ribbon tabs in Project shows different options on the Ribbon. In this section, I provide an overview of each Ribbon tab and of the Ribbon and the Quick Access toolbar. I elaborate on various functions and commands on the Ribbon in later chapters.

Each Ribbon tab has a different group of controls or functions. You can navigate from one tab to another by clicking on the tab name.

The first tab on the left is the File Ribbon tab. After you click this tab, you see the Backstage view with the Navigation pane down the left side, as shown in Figure 1-5.

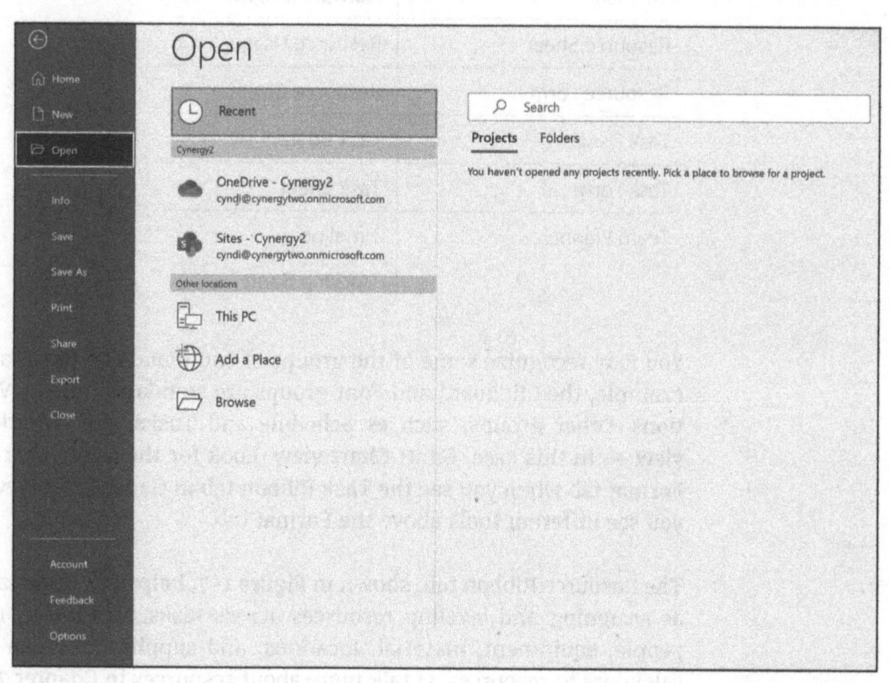


**FIGURE 1-5:** Backstage view with the Navigation pane.



The File tab puts you into Backstage view, where you find choices for working with files and changing options. For example, you can create a new project, open an existing project, save your current project, or print your current project. From Backstage view, you can also share, export, or close your current project. If you're feeling adventurous, you can click Options and customize the Ribbon and the Quick Access toolbar.

The Task Ribbon tab is where you spend a lot of your time in Project. As you can see in Figure 1-6, on the far left side of the Task Ribbon tab is the View group. The default view is Gantt Chart view. It shows the task information and the chart that displays a bar chart representing the duration of each task.

**FIGURE 1-6:** The Task Ribbon tab.

© John Wiley & Sons, Inc.

In addition to Gantt Chart, you can choose these views:

| | |
|---|---|
| Calendar | Network Diagram |
| Resource Sheet | Resource Usage |
| Resource Form | Resource Graph |
| Task Usage | Task Board |
| Task Form | Task Sheet |
| Team Planner | Timeline |
| | Tracking Gantt |

You may recognize some of the groups of commands on the Task Ribbon tab. For example, the Clipboard and Font groups are standard in many Windows applications. Other groups, such as Schedule and Tasks, are specific to a particular view — in this case, Gantt Chart view. Look for the Gantt Chart Tools above the Format tab when you see the Task Ribbon tab in Gantt Chart view. In other views, you see different tools above the Format tab.

The Resource Ribbon tab, shown in Figure 1-7, helps you organize resources, such as assigning and leveling resources across tasks. In Project, resources include people, equipment, material, locations, and supplies. You can assign costs and calendars to resources. (I talk more about resources in Chapter 7.)

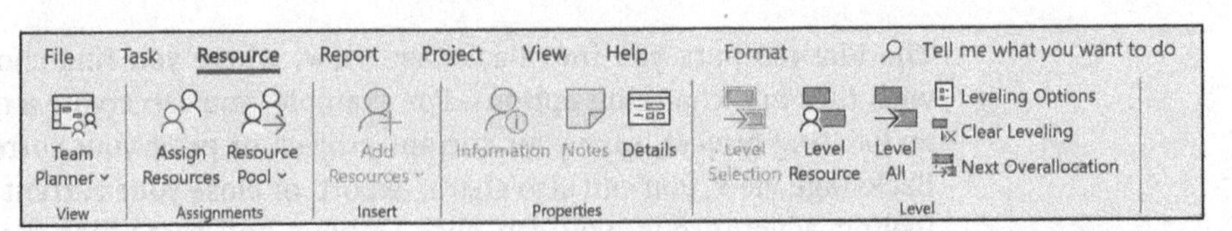


**FIGURE 1-7:** The Resource Ribbon tab.

© John Wiley & Sons, Inc.

The Report Ribbon tab, shown in Figure 1-8, is where you can create reports on resources, costs, or progress, or put them all together in a dashboard report. You can create a report that compares your current status to previous versions of your project. I tell you all about reports, including how to customize and export your reports, in Chapter 18.

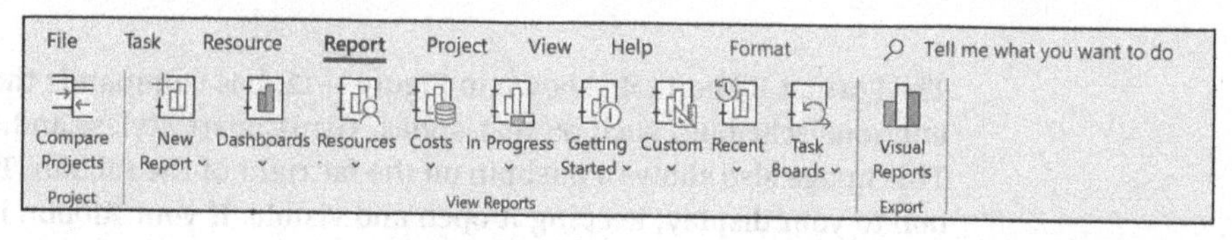


**FIGURE 1-8:** The Report Ribbon tab.

© John Wiley & Sons, Inc.

On the Project Ribbon tab, shown in Figure 1-9, you find commands to help you manage your project as a whole, rather than by task or resource. For example, you can enter or change the project start and finish dates and the baseline. If you need to change working time or add a subproject, this is the place to do it. You can also manage sprints from this Ribbon tab.

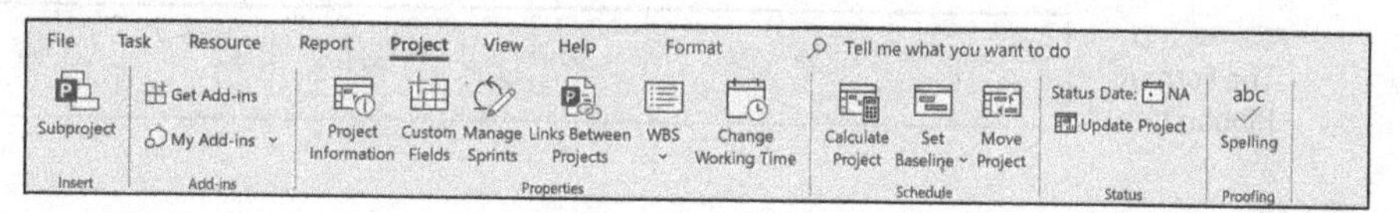


**FIGURE 1-9:** The Project Ribbon tab.

© John Wiley & Sons, Inc.

The View Ribbon tab, shown in Figure 1-10, lets you see some standard views. Examples are Task views, such as Gantt Chart, Task Usage, and Task Board. You can also check out Resource views, such as Resource Sheet or Team Planner. You can use the View Ribbon tab to look at information sorted by date or a specific period. This tab also lets you see the entire project, show or hide the Timeline, and set the timescale you see.

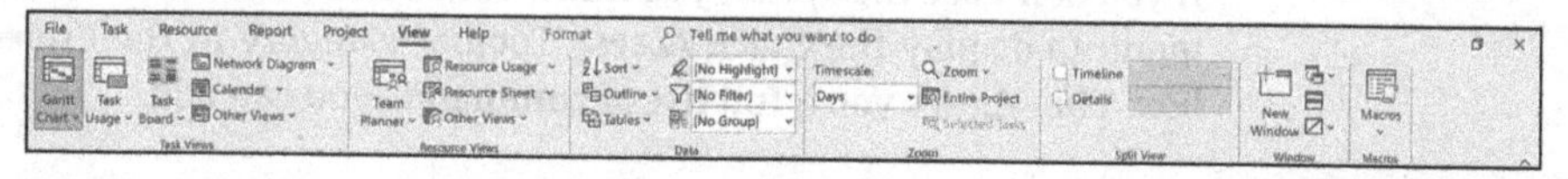


**FIGURE 1-10:** The View Ribbon tab.

© John Wiley & Sons, Inc.

The Help Ribbon tab, shown in Figure 1-11, lets you open a Help pane, provide feedback, and access training. It has a pane that highlights what's new. The Help function in the Help Ribbon tab opens a pane on the right side of the window. You can enter keywords and you will get several explanations with hyperlinks. Find the topic that reflects what you are looking for and click the link. Information in the Help pane often comes with a graphic and step-by-step instructions you can follow.

FIGURE 1-11: The Help Ribbon tab.

© John Wiley & Sons, Inc.

The Format Ribbon tab, shown in Figure 1-12, has commands that help you present your schedule, such as text styles, Gantt chart styles, and column settings. This image also shows a pushpin on the far right of the Ribbon. This pins the Ribbon to your display, keeping it open and visible. If your Ribbon is closed, click on any tab and look in the lower-right corner of the Ribbon to see the pushpin. Click on the pushpin to keep your Ribbon open.

The upward-facing arrow (∧) on the far-right side of a Ribbon (as shown in Figure 1-10) hides the Ribbon. This gives you more real estate on your screen.

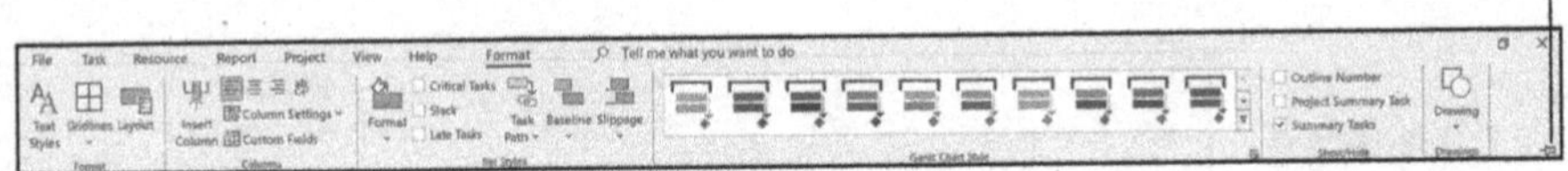


FIGURE 1-12: The Format Ribbon tab.

© John Wiley & Sons, Inc.

## Displaying more tools

The Quick Access toolbar, which appears onscreen at all times, initially contains the Save, Undo, and Redo buttons. You can customize the Quick Access toolbar by clicking the down arrow at the right end of the toolbar and clicking the option you want to hide or display. Figure 1-13 shows the list of commands you can choose from.

If you don't see the option you want, click More Commands near the bottom of the menu to display the Quick Access Toolbar category in the Project Options dialog box. This shows you a full list of commands you can add.

The nifty *Timeline* tool shows the entire scaled time span of the project. To show the Timeline, go to the View Ribbon tab (shown in Figure 1-10), locate the Split View group, and then click the check box that says Timeline. You can add tasks or milestones to the Timeline. You can also copy the Timeline and paste it into reports or other presentations. To hide the Timeline, uncheck the Timeline box. You can also work with the Timeline by right-clicking to insert tasks, copy the Timeline, change the font, or view detailed information. Figure 1-14 shows the Timeline with summary tasks and milestones.

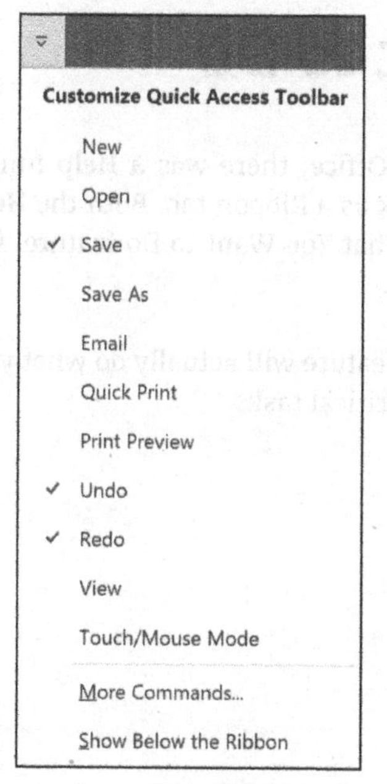


**FIGURE 1-13:** Customizing the Quick Access toolbar.

*© John Wiley & Sons, Inc.*

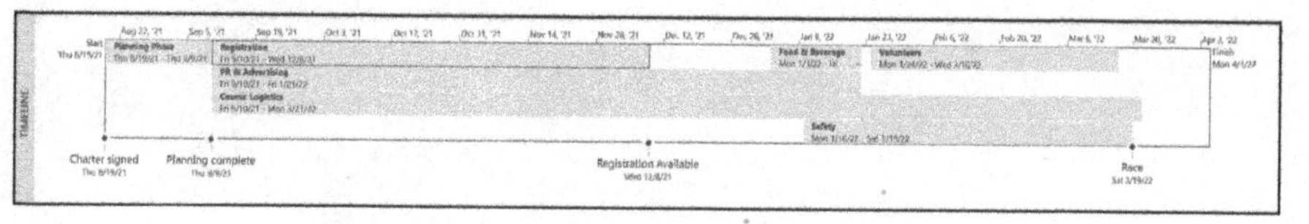


**FIGURE 1-14:** The Timeline.

*© John Wiley & Sons, Inc.*

The status bar, shown in Figure 1-15, sits at the bottom of the project, to indicate whether your tasks are manually or automatically scheduled. (Read more on this topic in Chapter 2.) The status bar also lets you move quickly to some of the most popular views, such as Gantt, Task Usage, Team Planner, Resource Sheet, and Reports. You can also adjust the time scale from a high-level, time scaled view to a detailed time-scaled view with the View slider, on the far-right end of the status bar. I talk more about views in Chapter 6.

**FIGURE 1-15:** The status bar.

*© John Wiley & Sons, Inc.*

# Tell Me What You Want to Do

In previous versions of Microsoft Office, there was a Help function. That went away for a while, but now it is back as a Ribbon tab. Both the Help button on the Help Ribbon tab and the Tell Me What You Want to Do feature, found next to the Format tab, can provide assistance.

TIP

The Tell Me What You Want to Do feature will actually do what you ask it to, such as insert a milestone or highlight critical tasks.

IN THIS CHAPTER

» Chartering a project

» Creating the project's work breakdown structure (WBS)

» Entering project information into Project

» Entering the WBS into Project

» Entering tasks into Project

» Inserting subprojects and hyperlinks

# Chapter 2
# Starting the Project

Congratulations — you're the proud project manager of a new project! Before you do anything, you need to understand the scope of the project. You must clearly specify what's in and out of scope, milestones, the budget, and completion criteria.

Planning a project isn't as easy as opening a file in Project and entering all the activities you have to complete. Before you can begin to plan, you need to understand the purpose of the project and the high-level information about the project such as project's objectives and the intended outcomes.

Therefore, before you enter the first task into Project, you create the project charter (described a little later in this chapter) to initiate the project and develop the work breakdown structure (WBS) to organize project deliverables. Then you can start organizing the project and entering tasks into Project.

This chapter describes how to move from imagining a concept to planning a project so you know how to enter and work with tasks and how to save the new project.

## Creating the Project Charter

Having a high-level understanding of a project is critical to project success. The *project charter* is a document that formally authorizes or recognizes a project; it contains high-level information about the project. The project charter is frequently developed by the project manager, with the project sponsor. The charter functions as an agreement about the purpose and objectives of the project.

In project management parlance, the person who champions (and funds) a project is the *project sponsor*. Although the project manager may work for the project sponsor, the project often also has a *customer* — outside the project manager's own company or within it — for whom the end product is produced.

Common elements of a project charter are:

- Purpose
- Description
- Objectives
- Criteria for completion
- Summary milestone schedule
- Summary budget

Other names for the charter are *project-initiating document* and *statement of work*.

The high-level information in the charter provides background information to help you plan the project approach and organize the work logically. Using the information from the charter, you can start to define the project's major deliverables and its life cycle — and your approach to accomplishing all the project work.

Throughout this book, I use a project to demonstrate key concepts in Project. The project is part of a larger program to build a community called *Desert Rose*. It is a gated community that will have four neighborhoods and community spaces and activities. Our project is the Security for the Desert Rose community. The project charter for the Security sample project is shown in Figure 2-1.

**Purpose**
Support the Desert Rose Community Program by providing a safe and secure environment.

**Description**
Provide a safe physical environment for Desert Rose residents by establishing a secure perimeter and points of entry. Hire and train qualified personnel and ensure they have the equipment needed to perform their jobs. Identify, purchase and manage the equipment and services needed to provide security 24/7 throughout the Desert Rose Community. Prepare for operations by hiring and training personnel.

**In Scope**

1. Design and construction of the community perimeter and all points of entry, including guard houses.
2. Identification and purchase of all necessary transportation equipment (trucks, carts, etc.).
3. Identification and purchase of all necessary communication equipment (radios, phones, tablets, etc.).
4. Identification and purchase of a community security system.
5. Work with Information Services to develop an asset management system for all physical assets in the community.
6. Work with HR to hire and train staff.
7. Work with HR to identify and develop security operations policies and procedures.

**Out of Scope**

- Operations beyond initial training.
- Maintenance and operations.
- Self-defense and first-aid training.
- Security systems for individual homes.

**Completion Criteria**

1. The community is fully fenced and gated with guard houses at the 4 main entry points.
2. The community has sufficient security trucks and carts to cover the community.
3. There are sufficient radios, cell phones and tablets to maintain communication with security base, local law enforcement, fire and paramedics.
4. A security system that covers all points of entry and common area buildings.
5. All security positions filled with qualified personnel.
6. A security operations manual that covers all policies, procedures, and operations information needed to operate Security for the community.

**Summary Milestones**

- ♦ Perimeter complete.
- ♦ Guard houses complete.
- ♦ Transportation equipment delivered.
- ♦ Communications equipment delivered.
- ♦ Security system installed and tested.
- ♦ Asset management system deployed.
- ♦ All staff hired.
- ♦ Operations manual complete.
- ♦ Staff training complete.

**Authorized Budget**
$5,000,000

**FIGURE 2-1:** The Desert Rose Security project charter.

# Introducing the Work Breakdown Structure (WBS)

If you have a small project, you may be able to start entering tasks into Project and organize them on the fly. But for any project with more than 50 tasks, consider how to structure and organize the work before you open Project. One best practice is to create the *work breakdown structure*, or WBS — a hierarchically organized representation of all the project work.

REMEMBER

The concept of *project work* includes work that's necessary to *create* the product and work that's necessary to *manage* the project, such as attending meetings, managing risk, and creating documentation.

Generally, you approach the WBS from the top down. In other words, you evaluate the entire project and then break it into large chunks, and then break the larger chunks into smaller chunks, and so on, until you have a defined deliverable. That's where the WBS stops and project tasks begin.

TIP

The breaking of WBS deliverables into smaller chunks is known as *decomposition*.

The WBS houses all deliverables for the project and product scope. It doesn't include the tasks. Those are strictly for the schedule. Another way of thinking about the WBS is that it's composed of nouns, whereas the schedule is composed of actionable verbs. For example, the Perimeter Fencing might be the lowest-level deliverable you would show on the WBS. Then define these tasks for the schedule using the "verb-noun" naming convention:

1. Identify fencing requirements.
2. Develop request for quote.
3. Receive quotes.
4. Select vendor.
5. Develop contract.
6. Sign contract.
7. Oversee fence installation.

# Organizing the Work

Frequently, the most challenging aspect of creating a WBS is figuring out how to organize it. You have several options. For example, if you have multiple locations for a hardware deployment, you can arrange it by geography.

Another common way to organize work is by major deliverable. The needs and deliverables of the project determine how best to organize work. The WBS for Desert Rose Security is arranged by the type of work. You can see it presented in two different ways. One way looks like an organizational chart, and the other way is via outline. Either way works well; however, when you start decomposing past two levels, the organizational chart method becomes difficult to manage. After you delve into the detail, consider moving to an outline format. Figure 2-2 shows an organizational chart format for the high-level WBS.

***Desert Rose Community Center WBS***

**Desert Rose Security Project**

1. Perimeter
   1.1. Entry gates
   1.2. Guard houses
   1.3. Walls
2. Equipment
   2.1. Trucks
   2.2. Carts
   2.3. Communications
   2.4. Alarm system
3. Asset Management
   3.1. Requirements
   3.2. Database
   3.3. Asset information
4. Operations Readiness
   4.1. Staffing
   4.2. Operations manual
   4.3. Training

**FIGURE 2-2:** High-level WBS.

*© John Wiley & Sons, Inc.*

Using the high-level WBS, you can further decompose the work into outline format:

**Desert Rose Community Center**

1. Perimeter
   - 1.1 Entry gates
   - 1.2 Walls
   - 1.3 Guard houses
2. Equipment
   - 2.1 Trucks
   - 2.2 Carts
   - 2.3 Communication
     - 2.3.1 Radios
     - 2.3.2 Phones
     - 2.3.3 Tablets
   - 2.4 Security system
3. Asset Management
   - 3.1 Requirements
   - 3.2 Database
   - 3.3 Asset information
4. Operations Readiness
   - 4.1 Staffing
   - 4.2 Operations manual
   - 4.3 Training

After outlining the organization of the work, you can start defining the tasks that comprise the project schedule.

# Starting the Project

In Chapter 1, I discuss how to open a blank project. After you open a blank project, you can begin entering basic project information, such as the start or end date.

# Entering project information

You can enter project information into Project in two ways:

- **Choose File ➪ Info.** In the Project Information section on the right side of the screen, as shown in Figure 2-3, you can enter the start, finish, schedule from, current, and status dates for the project. All you have to do to make an entry or change an entry is click on it to display the controls — for example, to change the date, just click on the date picker.

TIP

After choosing File ➪ Info, you can click the Project Information down arrow, then click Advanced Properties. The Properties dialog box opens. Enter the name of the project where it says "Title" and enter your name where it says "Author." You can also enter company name, keywords, and other information. This information is used in many of the automated reports that Project creates.

- **On the Ribbon, go to the Project tab and click the Project Information icon.** You see the Project Information dialog box, shown in Figure 2-4. It holds the same information as the Project Information section on the Info screen in the Backstage view.

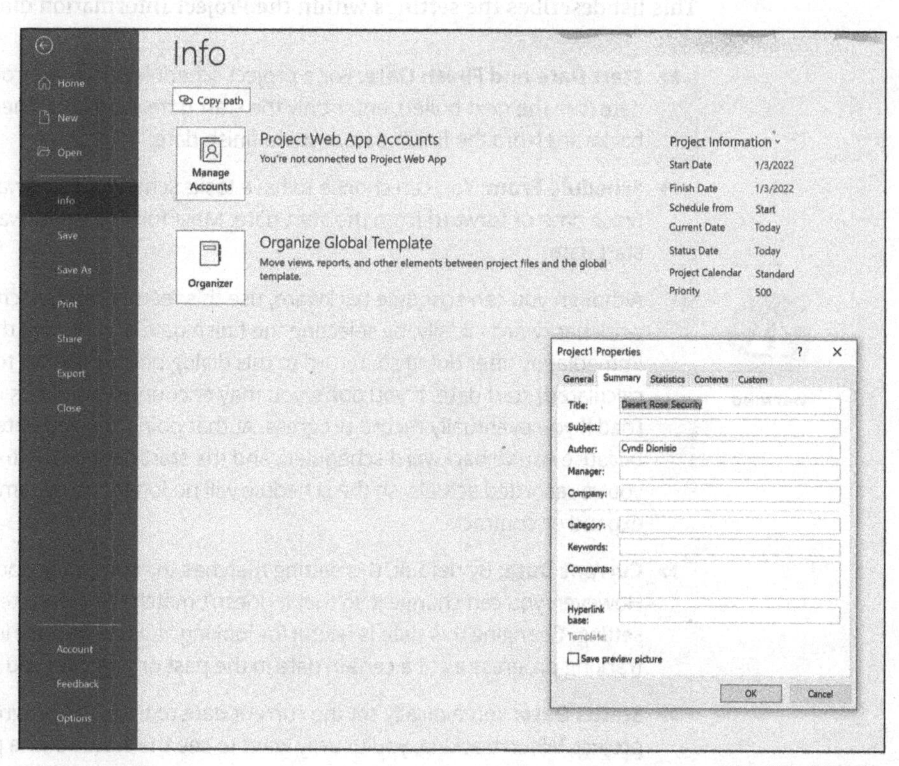


**FIGURE 2-3:** The result of choosing the File ➪ Info command.

© John Wiley & Sons, Inc.

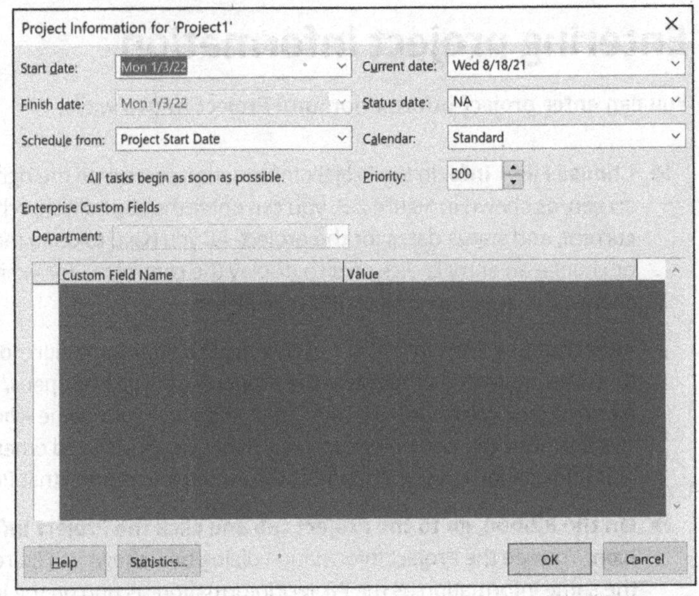


**FIGURE 2-4:** The Project Information dialog box.

© John Wiley & Sons, Inc.

This list describes the settings within the Project Information dialog box:

- **Start Date and Finish Date:** For a project scheduled forward from the start date (see the next bullet), enter only the start date. Or if you schedule backward from the finish date, enter a finish date.
- **Schedule From:** You can choose to have tasks scheduled backward from the finish date or forward from the start date. Most folks move forward from the start date.

WARNING

Although you can schedule backward, use this feature only to schedule the work backward initially, by selecting the finish date and clicking the OK button. Immediately after doing so, return to this dialog box and revert to the (newly calculated) start date. If you don't, you may encounter problems down the road if you eventually record progress. At that point, the end date will be fixed because you've backward-scheduled, and the start date will be fixed because you've recorded actuals, so the schedule will no longer be dynamic and won't expand or contract.

- **Current Date:** By default, this setting matches the computer clock setting. However, you can change it so that it doesn't match the computer clock setting. Changing this date is useful for looking at what-if scenarios or for tracking progress as of a certain date in the past or any date you choose.
- **Status Date:** You typically set the current date to track the progress on the project. When tracking, you usually want to see the status of the project as of

the current moment, so you can ignore this setting. However, if you want to track as of the end of a fiscal period or another timeframe, change it to track the status of tasks as of any other date.

- **Calendar:** For now, assume you will use the Standard calendar. I describe calendars in Chapter 8.
- **Priority:** This field is useful if your organization has many projects and you create links among them. If you use a tool such as resource leveling (see Chapter 12 for more about this topic) to resolve conflicts, it can consider this project priority setting when calculating what to delay and what to keep on track.

# Entering the WBS

The schedule should be set up the same way as the WBS. If you want to continue the outline numbering scheme from the WBS, Project automatically extends the outline numbering used for the WBS to the schedule tasks. You can do this in two ways:

1. **Select the column to the right of where you want to insert the new column.** I select the Task Name column because I like to insert this column to the left of the that column.
2. **Select the Format tab.**
3. **In the Columns group, click Insert Column.**
4. **Select the Outline Number option.**

This option uses a separate column to track your WBS and task numbering. You can also enter the outline number directly into the task cell by following these steps:

1. **Click to open the Format tab.**
2. **In the Show/Hide group, select the Outline Number check box.**

The first information to enter into the new project is the WBS. Follow these simple steps to enter the WBS in Gantt Chart view:

1. **In the Task Name column, click a blank cell.**
2. **Type the name of the WBS element.**

   Using the example of the **Desert Rose Security**, the first element is Perimeter.

3. **Press Enter to move to the next cell in the column and then type the WBS element name.**
4. **Repeat Step 3 until you enter all WBS names.**

   You can edit the text that you type by pressing the Delete or Backspace key to clear characters.

Figure 2-5 shows part of the screen display with the outline numbering after you enter the WBS outline and indent some of the tasks. You learn how to indent next.

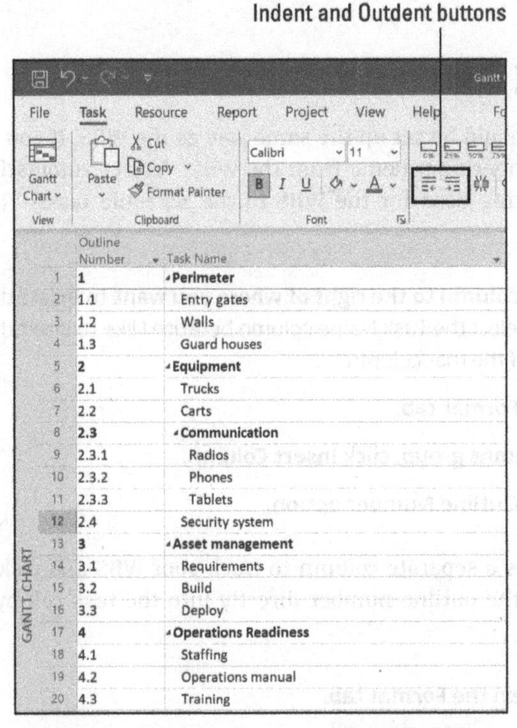


**FIGURE 2-5:** Desert Rose Security WBS.



## Indenting and outdenting (a.k.a. promoting and demoting)

Outdenting and indenting are the functions you use to move tasks to higher or lower levels of detail in the WBS and the task list. In several software programs, these terms are *promoting* and *demoting*, respectively:

* **Outdenting** a task moves it up a level in the outline (literally shifting it to the left in the outline).
* **Indenting** a task moves it down a level in the outline (literally indenting the task to the right in the outline). Whenever you indent one or more tasks, the task above becomes the summary task. You can read more about summary tasks in Chapter 3.

You use tools from the Schedule group on the Task tab of the Ribbon, shown in Figure 2-5, to outdent and indent tasks in a project outline. The Outdent Task tool has a left-facing arrow; the Indent Task tool has a right-facing arrow.

To outdent or indent a task, follow these steps:

1. **Click a task to select it.**
2. **Click the Indent Task or Outdent Task button, according to the action you want to take.**

   When you indent a task, the task above it becomes a summary task. The summary task is in bold on the sheet. On the chart, a summary task has a bracket that stretches from the beginning of the earliest task to the end of the latest task.

TIP

You can build the outline even faster by indenting multiple tasks at a time. Drag to select multiple task IDs and then indent them. You can also use the standard Shift+click and Ctrl+click selection methods to select multiple tasks in a Project outline. Shift+click the task IDs to select contiguous tasks, and Ctrl+click the IDs to select discrete tasks.

## Entering tasks

After the WBS information is entered, you can start entering tasks. You can create tasks in a couple different ways:

* Type information in the sheet area of the Gantt chart.
* Enter information in the Task Information dialog box.

You can fill in the details of the task duration and start date when you enter the task, or later.

## Entering tasks in Gantt Chart view

Many people who work on lengthy projects find that entering all task names in the sheet pane of Gantt Chart view is the quickest and easiest method. This method is the same one I used to enter the WBS information. You can simply enter a task name in the Task Name column, press the Enter or down-arrow key on the keyboard to move to the next blank row, enter another task, and so on.

## Entering tasks via the Task Information dialog box

If dialog boxes provide the kind of centralized information form that fits the way you like to work, consider using the Task Information dialog box. This box has a series of tabs that contain all the information about a task.

Follow these steps to create a task via the Task Information dialog box:

1. **In the Task Name column, double-click a blank cell.**

   The Task Information dialog box appears, as shown in Figure 2-6.

2. **In the Name field, type a task name.**

   You can enter any other information you would like to while you are there.

3. **Click the OK button to save the new task.**

   The task name appears in Gantt Chart view in the cell you clicked in Step 1.

4. **Repeat Steps 1–3 to add as many tasks as you like.**

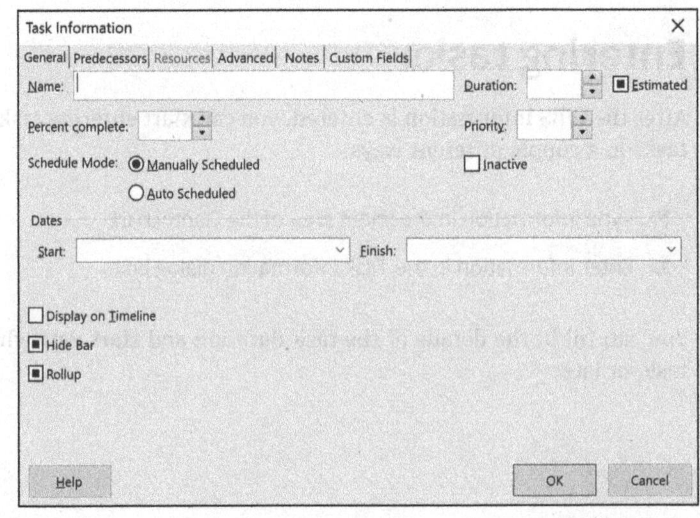


**FIGURE 2-6:** The Task Information dialog box.

© John Wiley & Sons, Inc.

As you name tasks, make task names in the project both descriptive and unique. However, if you can't make all names unique (for example, you have three tasks named Hire Staff), you can use the automatically assigned task number or the outline number to identify tasks; these numbers are always unique for each task.

REMEMBER

Naming tasks is a trade-off between giving a full description (which is much too long for a Task Name field) and being too brief (which can lead to misunderstandings and uncertainty). When in doubt, be brief in the Task Name and elaborate with a Task Note. I cover Task Notes in Chapter 3.

TIP

To insert a task anywhere within the list of tasks in Project, from the Task tab, Insert group, click a task name cell where you want the new task to appear, and click the Insert Task icon. The new task is inserted in the row above. You can also press the Insert key on the keyboard, or right-click and select Insert Task from the pop-up menu.

## Weighing manual scheduling versus automatic scheduling

One of the most valuable aspects offered by Project has traditionally been its ability to recalculate task schedules, such as when you change the project start date or there is a change to one task's schedule that affects one or more dependent *(linked)* tasks. This powerful behavior saves the project manager — *you* — from having to rethink and reenter dates to rescheduled tasks throughout the project.

But flip sides to the benefits of automation always exist, and in the case of project scheduling, automatic scheduling can lead to unwanted schedule changes based on software behavior and not on human expertise.

To retain the helpful aspects of automation that make scheduling less time-consuming while allowing project managers to retain schedule control when needed, Project allows *user-controlled scheduling.*

In user-controlled scheduling, you can select one of these scheduling modes for each task:

- **Auto Scheduled:** Project calculates task schedules for you based on the project start date and finish date, task dependencies, calendar selections, and resource scheduling.
- **Manually Scheduled:** Project enables you to skip entering the duration and dates, and specify them later. When you enter the duration and dates, Project fixes the schedule for the task and doesn't move it unless you do so manually.

The manually scheduled tasks move if you reschedule the entire project, in most cases. The Gantt bars for manually scheduled tasks also differ in appearance from those for automatically scheduled tasks.

The indicator for auto-scheduled and manually scheduled tasks is at the bottom of the Project window. Figure 2-7 shows that the Walls tasks are auto-scheduled, as indicated by the time bar and the arrow in the Task Mode column. The Entry gates are manually scheduled as indicated by the pushpin in the Task Mode column. On the time scale, the auto-scheduled tasks show up as blue bars on your screen and the manually scheduled tasks show up as aqua bars with vertical lines on each end.

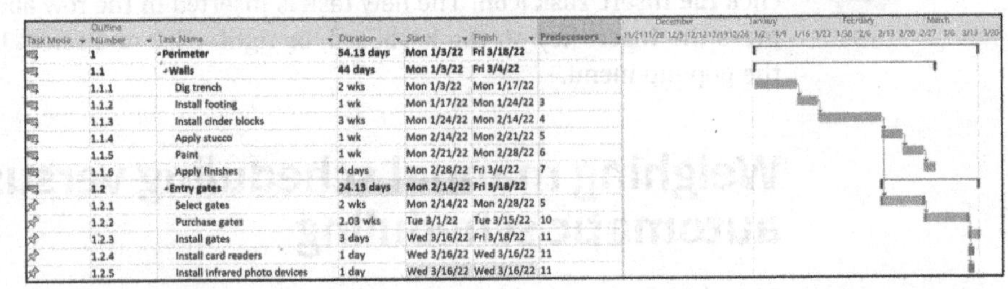

| Task Mode | Outline Number | Task Name | Duration | Start | Finish | Predecessors |
|---|---|---|---|---|---|---|
| | 1 | Perimeter | 54.13 days | Mon 1/3/22 | Fri 3/18/22 | |
| | 1.1 | Walls | 44 days | Mon 1/3/22 | Fri 3/4/22 | |
| | 1.1.1 | Dig trench | 2 wks | Mon 1/3/22 | Mon 1/17/22 | |
| | 1.1.2 | Install footing | 1 wk | Mon 1/17/22 | Mon 1/24/22 | 3 |
| | 1.1.3 | Install cinder blocks | 3 wks | Mon 1/24/22 | Mon 2/14/22 | 4 |
| | 1.1.4 | Apply stucco | 1 wk | Mon 2/14/22 | Mon 2/21/22 | 5 |
| | 1.1.5 | Paint | 1 wk | Mon 2/21/22 | Mon 2/28/22 | 6 |
| | 1.1.6 | Apply finishes | 4 days | Mon 2/28/22 | Fri 3/4/22 | 7 |
| | 1.2 | Entry gates | 24.13 days | Mon 2/14/22 | Fri 3/18/22 | |
| | 1.2.1 | Select gates | 2 wks | Mon 2/14/22 | Mon 2/28/22 | 5 |
| | 1.2.2 | Purchase gates | 2.03 wks | Tue 3/1/22 | Tue 3/15/22 | 10 |
| | 1.2.3 | Install gates | 3 days | Wed 3/16/22 | Fri 3/18/22 | 11 |
| | 1.2.4 | Install card readers | 1 day | Wed 3/16/22 | Wed 3/16/22 | 11 |
| | 1.2.5 | Install infrared photo devices | 1 day | Wed 3/16/22 | Wed 3/16/22 | 11 |

**FIGURE 2-7:** Manually and automatically scheduled tasks.



When you indent a task, its parent task switches from manually scheduled to auto-scheduled because the duration and dependencies of the child tasks determine when the parent task can start and finish. Therefore, you don't fill in the duration or start and finish dates for WBS elements — that information will auto-populate when you enter tasks beneath the WBS elements.

The project file can have all manually scheduled tasks or all auto-scheduled tasks — or any mix of the two. By default, all tasks that you create use the manually scheduled mode.

You can change the task mode for the overall project in two ways:

- To change the mode for all new tasks, select the Task tab, click the Mode icon in the Tasks group (the button with the calendar and a question mark), and then choose Auto Schedule or Manually Schedule from the menu, as shown in Figure 2-8.
- Another way to change the mode for all new tasks is to click the New Tasks link at the left end of the status bar at the bottom of the project, then select the mode you want.

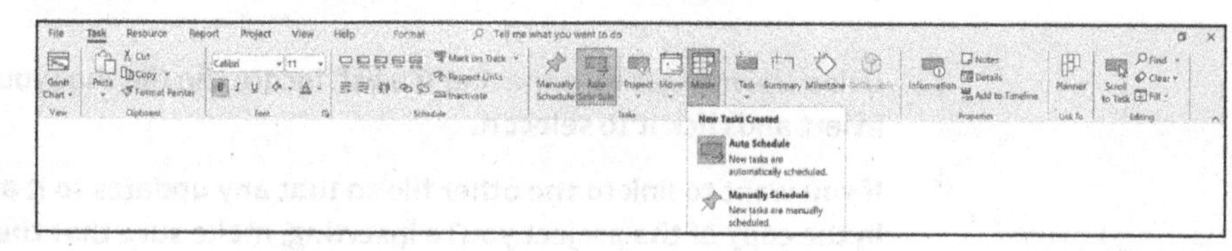


**FIGURE 2-8:** Changing the task mode from the Ribbon.

*© John Wiley & Sons, Inc.*

You can change the task mode for individual tasks in three ways:

- Select the task, click the Task tab on the Ribbon, and then click either Manually Schedule or Auto Schedule in the Tasks group.
- Select the task, click the Task Mode cell for the task, click the drop-down arrow that appears, and click either Manually Scheduled or Auto Scheduled in the drop-down list.
- Select the task, click the Task tab on the Ribbon, and click the Information button in the Properties group. On the General tab of the Task Information dialog box, go to the Schedule Mode area and click the Manually Scheduled radio button or the Auto Scheduled option button.

REMEMBER

You need to balance the desires of your inner control freak versus the need to be an efficient project manager in determining how often to use manual scheduling. Though manual scheduling prevents Project from moving tasks that you want to stay put in the schedule, you may need to edit the schedules for dozens of dependent tasks in a long or complicated project. The best balance — particularly for beginning project managers — may be to use manual scheduling sparingly.

## Inserting one project into another

You can also insert tasks from one project into another. You do this by inserting an entire, existing project into another project. The project that's inserted is called a *subproject.* This method is useful when various project team members manage different phases of a larger project. The capability to assemble subprojects in one place allows you to create a master schedule from which you can view, all in one place, all the pieces of a larger, more complex project.

Follow these steps to insert another Project file into the schedule:

1. **In Gantt Chart view, select the task in the task list above which you want the other project to be inserted.**
2. **From the Project tab, in the Insert group, select Subproject.**

   The Insert Project dialog box appears, as shown in Figure 2-9.

3. **Using the navigation pane and file list, locate the file that you want to insert and click it to select it.**

4. **If you want to link to the other file so that any updates to it are reflected in the copy of the project you're inserting, make sure that the Link to Project check box is selected.**

5. **Click the Insert button to insert the file.**

   The inserted project appears above the task you selected when you began the insert process. You may want to choose Insert Read-Only from the Insert drop-down list if you just want people to be able to view the file, but not make any changes to it.

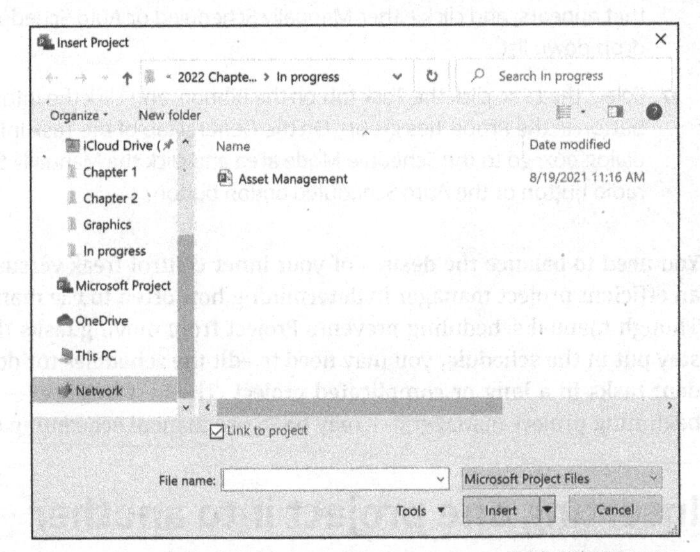


**FIGURE 2-9:** Inserting a link to another project.

© John Wiley & Sons, Inc.

REMEMBER

The inserted project's highest-level task appears at the level of the task you selected when you inserted the project, with all other tasks below it in outline order. If you need to, use the Outdent Task button and Indent Task button on the Formatting tab to place the inserted tasks at the appropriate level in the project.

## Inserting hyperlinks

You can insert hyperlinks in a project outline, which provides a handy way to quickly open another project, another file of any type, or a web page.

To insert a hyperlink to a project document, follow these steps:

1. **Right-click the cell where you want the hyperlinked task to appear.**
2. **Choose Link.**

   The Insert Hyperlink dialog box appears, as shown in Figure 2-10.
3. **In the Text to Display box, type the text that you want to appear for the hyperlink.**

   Ensure that this text clearly states what information is being summarized. In this case, I'm linking the Requirements Template to a task in the project.
4. **In the Link To area, click the Existing File or Web Page icon.**

   You can link to a document of any type or to a web page.
5. **In the Look In list, locate and select the file to which you want to insert a hyperlink.**
6. **Click the OK button.**

   The link text is inserted, and a hyperlink symbol appears in the Indicator field. You can simply click that link symbol to open the linked file, or right-click and choose Hyperlink, then choose Open Link, or Open in New Window.

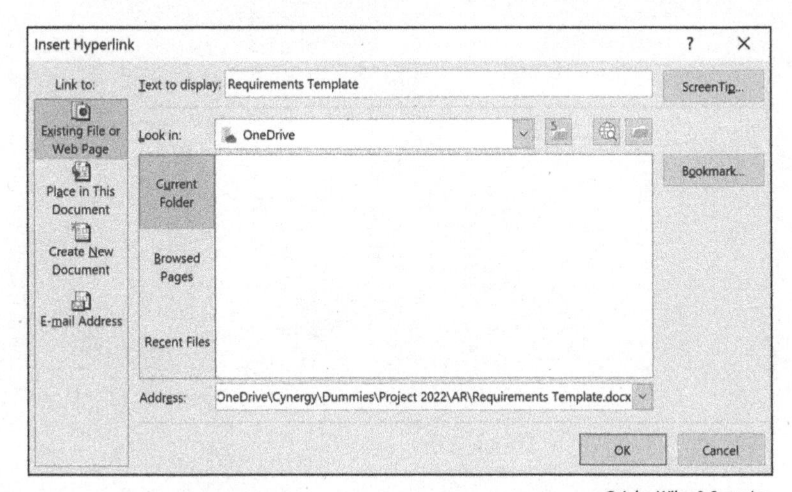


**FIGURE 2-10:** Linking a file.

*© John Wiley & Sons, Inc.*

TIP

Good organizational practice suggests that you create a folder for your project where you save your Project file, any linked files and supporting documents, such as the Charter, risk register, and other items. You can create a new folder from within the Save As dialog box by clicking the New Folder button.

IN THIS CHAPTER

» Creating summary tasks and subtasks

» Moving tasks

» Collapsing and expanding tasks

» Creating recurring tasks

» Creating milestones

» Deleting tasks and making tasks inactive

» Entering a task note

Chapter 3

# Becoming a Task Master

The foundational unit in a project schedule is the project task; everything starts with it. After you identify and document your tasks, you can work with resources, dependencies, costs, durations, and other elements.

To be an effective task master, you need to be nimble in managing your tasks: Determine how to summarize work with a summary task, move work around, start and stop work in the middle of a task, and do all kinds of other tricks that help your schedule reflect what you want to happen on your project.

## Creating Summary Tasks and Subtasks

When you look at a project work breakdown structure, also known as WBS (refer to Chapter 2), or a project task list, such as the one shown in Figure 3-1, you see that it organizes tasks into levels. The upper levels are from the WBS. The lower level consists of tasks that have been decomposed from the WBS. A task that has other tasks indented below it in this outline structure is a *summary task*, or *parent task*. The tasks indented below the summary task are sometimes known as *subtasks* or *child tasks*. Summary tasks are indicated in bold in the Project outline. You can tell when a summary task has subtasks even when they are hidden because a little clear triangle is displayed to the left of the summary task. When you click the triangle with the tip of the pointer, the task expands to show the list of subtasks, and the summary task adds a black triangle to its left.

- **Perimeter**
  - **Walls**
    - Dig trench
    - Install footing
    - Install cinder blocks
    - Apply stucco
    - Paint
    - Apply finishes
  - **Entry gates**
    - Select gates
    - Purchase gates
    - Install gates
    - Install card readers
    - Install infrared photo devices
  - Guard house
- **Equipment**
- **Asset Management**
- **Operations Readiness**

**FIGURE 3-1:** Summary tasks and subtasks.



In Figure 3-1 you can see that the Walls and Entry Gates summary tasks show all their subtasks. The summary tasks of Equipment, Asset Management, and Operations Readiness have hidden subtasks.

All information about a family of tasks is rolled up into its highest-level summary task. Therefore, any task with subtasks has no timing or cost information of its own: It gathers its total duration and cost from the sum of its parts.

This roll-up functionality is cumulative: The lowest-level task rolls up to its parent, which might roll up into another summary task, which rolls up (for example) into a project summary task. Any task with tasks below it gets its duration and cost information from a roll-up of its subtasks, no matter how deeply nested it may be in the hierarchy.

TIP

When you need to reorganize an outline, you can move a summary task and all its subtasks come with it, regardless of whether it's expanded.

If a summary task is manually scheduled, the roll-up functionality doesn't work, and Project displays warnings telling you so. The Gantt bar for the summary task has a red warning bar that shows you the calculated duration of the subtasks when they don't match up with the summary task's duration. The summary task's Finish field entry also has a red, squiggly underline to indicate a potential scheduling problem. To deal with this situation, you can change the summary task to use auto-scheduling, in which case it calculates roll-up data correctly. If you want the summary task to continue to be manually scheduled, you can edit its finish date or use the Task Inspector, as described in Chapter 11, to fix the summary task schedule.

## How many levels can you go?

You have no practical limit on how many levels of tasks you can create in an outline. Project enables you to indent to more levels of detail than you'll need for all but the most complex schedules. Remember, though: At some point, you have to deal with assigning timing and resources to each of these tasks and then track their progress. Too much detail can make your schedule difficult to manage. For example, if your project is a few months long, you don't want to track to a level where tasks last only a few hours. Best practices suggest that you always set up your schedule to the level to which you want to manage your team — typically, business (working) days or weeks.

For longer projects, you can schedule by using *rolling wave planning,* a method of progressively elaborating the amount of detail for near-term work and keeping at a higher level any work that's further out. For example, if you have a two-year project, you may have the first three months planned out in detail, the next three months at a higher level, and the remainder of the project schedule showing only milestones and key deliverables. As you progress through the project, you start to add more detail for six months and beyond. A good rule of thumb is to keep a good amount of detail for 90 days out.

WARNING

Rolling wave planning isn't an excuse to add scope; it's only the elaboration of existing scope.

## The project summary task

A *project summary task* represents the highest (least detailed) level of information and is often simply the title of the project, such as New Product Rollout. When you tell Project to display the project summary task, every task in the project falls under it in the outline, as shown in Figure 3-2.

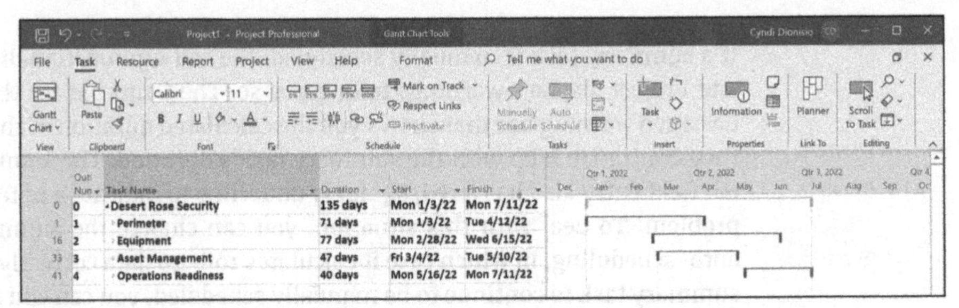

FIGURE 3-2: The project summary task.

© John Wiley & Sons, Inc.

TIP

The summary task rolls up all data from other tasks into one line item. Thus, the project summary task's duration reflects the duration of the entire project. From a monetary angle, the project summary task's total cost reflects the total costs for the entire project. Figures such as these can be handy to have at your fingertips — and that's one value of a summary task.

As you build your project, you can easily create a project summary task yourself (indent other tasks beneath it) or use a Project feature to generate one automatically at any time — even after you build all the phases of your project. To have Project automatically display a project summary task, follow these steps:

1. **In the Gantt Chart view, select the Format tab in the Gantt Chart Tools context tab group.**
2. **In the Show/Hide group, select the Project Summary Task check box.**

As you can see in Figure 3-2, Task 0, Desert Rose Security, is the project summary task. Notice the bar for the project summary task on the chart is gray and that the bars for other summary tasks are black.

REMEMBER

If you're confused about the length of your summary task, remember that the *summary task duration* is the difference between the earliest task start date and latest task end date. However, nonworking days aren't counted in the summary task duration. The length of the summary task, therefore, equals the number of days of work over the course of the subtasks, not the number of calendar days between the start of the first task and end of the last.

Not everyone uses project summary tasks. You can simply create, at the highest level of your outline, tasks that represent major project deliverables or phases — with subphases and subtasks below them — and not create one task that's higher in the order than all others. However, having a project summary task has certain benefits:

» **You can quickly view totals for the project at a glance in the columns of data in Gantt Chart view and other views.**

» **You can place a link to your project summary task in another project so that all data for one project is reflected in another.** For example, Desert Rose Security is one project in the Desert Rose Community Program. If there are projects for each of the four neighborhoods in the community, you can create one schedule for security and one for each of the four neighborhoods. Then you can easily create a master schedule for the whole program by linking to the project summary tasks in each of the projects.

# Moving Tasks Up, Down, and All Around

In Chapter 2, I show you how to outdent and indent tasks to create the WBS and the project outline. In this section, I show you additional ways to move tasks as well as collapse and expand tasks.

A maxim of project management says that things change: Tasks that you thought you could complete early can't happen yet because money, people, or materials are in short supply. Or a task that you thought you couldn't start until next July gets bumped up in priority when your customer changes their mind (again) about deliverables. Because of this changeability, when you enter tasks in a project outline, odds are that you'll need to move those tasks around at some point. You should understand that moving a task can change its outline level, so you'll need to check your outline level for tasks that you move and adjust it as necessary using the indent and outdent features.

## Moving tasks with the drag-and-drop method

If you ask me, drag-and-drop is to computing what the remote control is to television. It's a quick, no-brainer method of moving stuff around in software that just makes life simpler. Here's an example:

To move a task up and down with the drag-and-drop method, follow these steps:

1. **Display a column view, such as Gantt Chart view.**
2. **Select a task by clicking its task ID number.**

   Simply click and release; don't hold down the mouse button.

3. **Click and drag the task to wherever you want it to appear in the outline.**

   A gray "T-bar" line appears, indicating the new task position.

4. **When the gray line is located where you want to insert the task, release the mouse button.**

   The task appears in its new location. If you want the task to be at a different level of the outline, you can now indent or outdent it as needed.

REMEMBER

If you move the parent task, all the child tasks, and all the relevant information — such as start, finish, duration, cost, and so on — come with it.

## Moving tasks with the cut-and-paste method

Dragging and dropping works fine in most cases, but in very large projects — with a few hundred tasks or more, for example — it's easier to cut and paste instead of scrolling through hundreds of tasks to find the exact location you want to move your tasks to.

In a larger outline, it's best to use the cut-and-paste method to move tasks:

1. **Select a task by clicking its task ID number.**
2. **Click the Cut button in the Clipboard group on the Task tab.**

   The task is removed from its current location and placed on the Windows Clipboard.

3. **Scroll to display the location where you want the task to appear.**
4. **Click the task after which you want to insert the task.**
5. **Click the top part of the Paste button, also on the Task tab.**

If you want to insert a copy of a task in a project outline, you can follow the preceding steps and click Copy rather than Cut.

TIP

If you're cutting and copying only a single cell and not a whole task, click in the cell rather than clicking the task ID number.

TIP

You can also use the standard Microsoft shortcuts — Ctrl+C to copy, Ctrl+X to cut, and Ctrl+V to paste.

# Now You See It, Now You Don't: Collapsing and Expanding the Task Outline

Remember in elementary school when your teacher made you create an outline to organize your work? The outline helped you arrange content and allowed you to focus on different levels of information to keep everything organized.

With the invention of computer outlining, the capability to focus on only certain portions of an outline comes into its own, because you can easily expand and collapse an outline to show or hide different levels of information — or entire sections of your outline. The black triangle symbol next to the Walls summary task in Figure 3-1 indicates that all subtasks below it are displayed. The clear triangle next to the Equipment summary task indicates that the subtasks aren't displayed. Remember that all summary tasks are in bold in the project outline.

This capability means that you can hide all but the upper level of tasks in a project to give your manager an overview of progress. Or you can collapse every phase of your project except the one in progress so that your team can focus on only those tasks in a status meeting. Or you can collapse most of your outline to make it easy to move to a late phase of a very large schedule.

If you want to hide all the summary tasks and just see the child tasks, you can go to the Format tab in the Gantt Chart Tools context tab group and uncheck the Summary Tasks box (see Figure 3-2). Another nifty feature in Project is the ability to choose which level of detail you want to see in your outline. To wrangle the tasks to the level you want them, it's best to work with the View tab. In the View tab under the Data group, there is a drop-down option called Outline. Figure 3-3 shows you the commands you can use to control how your outline appears.

You can use the Show Subtasks command and the Hide Subtasks command at any level of summary tasks. Just select the summary task you want — regardless of whether it's at Level 1 or Level 5 — and click Show Subtasks to see everything underneath, or click Hide Subtasks to just see the summary task.

When you want an overview of your project, it's helpful to look at it from Level 1 or Level 2. You can see Desert Rose Security at Level 1 in Figure 3-2. Notice the duration for each of the summary level tasks. You can see how they nest within the Project Summary Task.

Figure 3-4 shows Perimeter and Equipment at Level 2. This level gives you a good overview of the work with a bit more detail than Level 1. Level 2 is the view I like to show management or other high-level stakeholders to give them an overall understanding of the project.

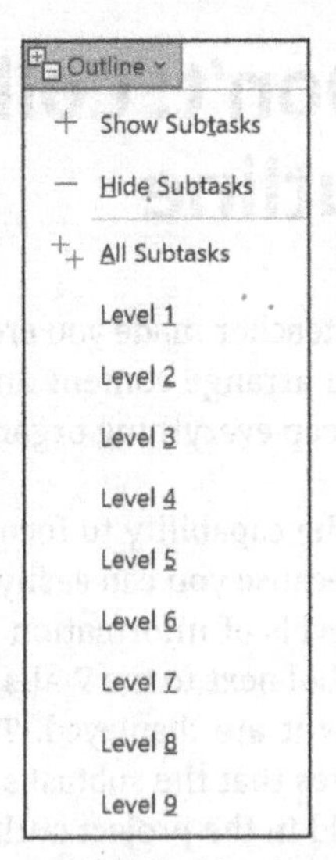


**FIGURE 3-3:** Outline options.

© John Wiley & Sons, Inc.

| Outl Num | Task Name | Duration | Start | Finish | Predecess |
|---|---|---|---|---|---|
| 0 | Desert Rose Security | 135 days | Mon 1/3/22 | Mon 7/11/22 | |
| 1 | Perimeter | 53 days | Mon 1/3/22 | Thu 3/17/22 | |
| 1.1 | Walls | 44 days | Mon 1/3/22 | Fri 3/4/22 | |
| 1.2 | Entry gates | 23 days | Mon 2/14/22 | Thu 3/17/22 | |
| 2 | Equipment | 117 days | Mon 1/3/22 | Wed 6/15/22 | |
| 2.1 | Communication | 15 days | Mon 1/3/22 | Mon 1/24/22 | |
| 2.2 | Security system | 77 days | Mon 2/28/22 | Wed 6/15/22 | |

**FIGURE 3-4:** Level 2 outline.

© John Wiley & Sons, Inc.

TIP

**To quickly reveal all subtasks in a project, click the Outline button and then click All Subtasks.**

TIP

**To show subtasks from the keyboard, you can press Alt+Shift+* (asterisk). To hide subtasks, you can press Alt+Shift+– (hyphen) or Alt+Shift+– (minus sign) if you're using the numeric keypad.**

# Showing Up Again and Again: Recurring Tasks

**Some tasks occur repeatedly in projects. For example, attending a monthly project debriefing or generating a quarterly project report is considered a recurring task.**

No one wants to create all the tasks for the monthly debriefing in a project that will take a year to complete. Instead, you can designate the recurrence, and Project automatically creates the 12 tasks for you.

Here's how to create a recurring task:

1. **Click the Task tab on the Ribbon, click the bottom part of the Task button (with the down arrow) in the Insert group, and then click Recurring Task.**

   Figure 3-5 shows the Recurring Task Information dialog box that will insert a recurring task for a monthly meeting on the 10th of every month. The meeting is scheduled to occur every month from January through July.

2. **In the Task Name box, type a name for the task.**
3. **In the Duration box, click the spinner arrows to set a duration, or type a duration, such as 1d for 1 day.**

   Edit the abbreviation to specify a different time unit, if needed.

   TIP

   You can read about the abbreviations that you can use for units of duration — such as *h* for hours — in Chapter 5.

4. **Select a recurrence pattern by selecting the Daily, Weekly, Monthly, or Yearly radio button.**

   The option you select provides different choices for the rest of the recurrence pattern.

5. **Depending on the selections offered to you, make choices for the rest of the pattern.**

   For example, if you select the Weekly radio button, you must choose a Recur Every *x* Week(s) On setting and then choose a day such as Friday. Or, if you select Monthly, you must specify every two months, every three months, and so on. You also need to choose which day of the month the task is to recur.

6. **In the Range of Recurrence area, type a date in the Start box; then select and fill in either the End After or End By option.**

   For example, you might start on January 1 and end after 12 occurrences to create a task that occurs every month for a year.

7. **Click the OK button to save the recurring task.**

If your settings cause a task to fall on a nonworking day (for example, if you choose to meet on the eighth day of every month and the eighth day is a Sunday in one of those months), a dialog box appears, asking you how to handle this situation. You can choose not to create the task, or you can let Project adjust the day to the next working day in that period.

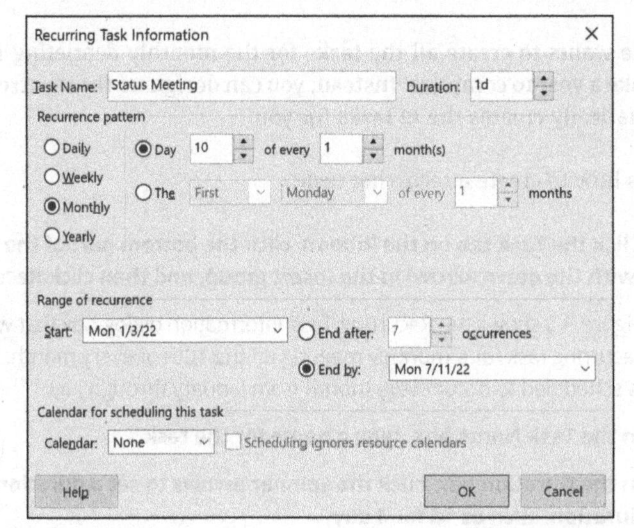


**FIGURE 3-5:** The Recurring Task Information dialog box.

© John Wiley & Sons, Inc.

TIP

To assign resources to a recurring task, you have to assign the resources to the individual recurrences, not to the "summary" recurring task. If you assign resources to the summary recurring task, Project doesn't calculate the hours of work correctly. I talk about assigning resources to tasks in Chapter 9.

# Setting Milestones

*Milestones* are signposts that mark significant events. Examples of milestones are the approval of a prototype (although the deliberations to make that decision might have taken months), the completion of a key deliverable, or the start or end of a project phase.

Some people include tasks such as Design Completed or Testing Completed at the end of each phase of their projects. They can then create timing relationships to the moment of completion — for example, allowing the production of a drug to proceed after the testing and approval is complete. Such milestones also alert you and your team members to a moment of progress in your project that can help keep the team motivated.

Project creates new tasks with an estimated duration of one day unless you enter a duration. To create a milestone, you indicate that the task has zero duration. There are several ways to create a milestone:

» An easy way to do this is to simply type **0** in the Duration column in Gantt Chart view. When you do, the milestone is designated in Gantt Chart view with a diamond shape rather than a taskbar, as shown in Figure 3-6.

» You can also click Milestone in the Insert group on the Task tab of the Ribbon. A new task will show up that says <New Milestone>.

» Finally, you can click the Advanced tab in the Task Information dialog box and select the Mark Task As Milestone check box. Use this last method for any milestone that has a duration other than zero but that you want to mark as a milestone anyway; its milestone marker is charted at the end of the duration period on the Gantt chart.

| | | |
|---|---|---|
| 1.1 | ◢ **Walls** | **44 days** |
| 1.1.1 | Dig trench | 2 wks |
| 1.1.2 | Install footing | 1 wk |
| 1.1.3 | Install cinder blocks | 3 wks |
| 1.1.4 | Apply stucco | 1 wk |
| 1.1.5 | Paint | 1 wk |
| 1.1.6 | Apply finishes | 4 days |
| 1.1.7 | Walls complete | 0 days |

3/4

**FIGURE 3-6:** Inserting a milestone.



# Deleting Tasks and Using Inactive Tasks

If your style is to delve into a lot of planning detail for everything you do, you may tend to overanalyze and add more tasks than you need to track in a project plan. In such cases, when you're refining the project plan, you may decide that you need to delete a task or two to tighten up the plan. After all, the process of planning includes not only decomposing the work, but also consolidating tasks when they become too detailed. To delete a task, click its row (Task ID) number at the left side of the sheet and press the Delete key.

WARNING

Project doesn't display a warning or ask you to confirm deleting a task, so make sure that you *really* mean it when you press Delete. If you screw up, click the undo arrow in your Quick Access toolbar or press Ctrl+Z immediately to restore the deleted task. Also, be sure to check that any task dependencies (links) adjust as needed after you delete the task.

Another approach is to mark a task as *inactive*. Doing this leaves the task visible in the plan but strikes it out in the task sheet and on the Gantt chart, as shown in Figure 3-7. If any automatically scheduled tasks were dependent on the inactive task, Project ignores the now-inactive task when calculating the schedules for those other tasks. In the example in Figure 3-7, the milestone of Perimeter Complete moved from April 12 to March 15 due to inactivating the Guard House task.

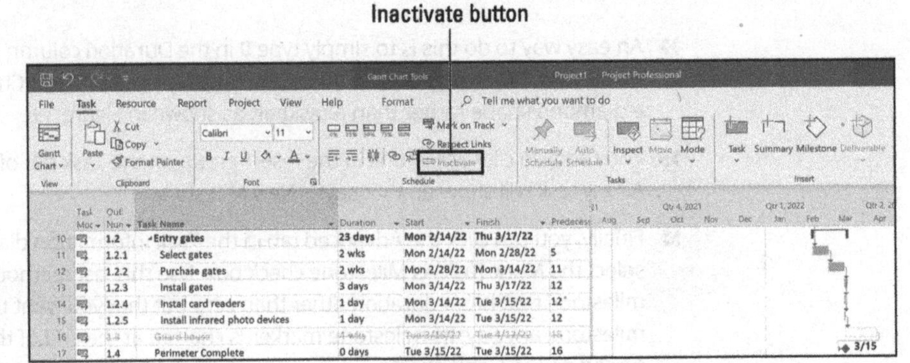


**FIGURE 3-7:** Marking a task as inactive.

*© John Wiley & Sons, Inc.*

Follow these steps to mark a task as inactive:

1. **In Gantt Chart view, select the task to make inactive.**
2. **Click the Inactivate button in the Schedule group on the Task tab of the Ribbon.**

   Project immediately reformats the task as inactive.

Leaving an inactive task in the schedule provides another opportunity to track and document what happened. You can do this by adding a task note explaining why the task was removed.

# Making a Task Note

Despite the wealth of information that you can enter about a task and its timing, not everything can be said with settings. That's why every task includes an area to enter notes. You may use this feature, for example, to enter background information about constraints, detail a step-by-step process that's summarized by the task duration, enter information about stakeholders, or list vendor contact information that's relevant to the task.

To enter task notes, follow these steps:

1. **Double-click a task.**

   The Task Information dialog box appears.

2. **Select the Notes tab, as shown in Figure 3-8.**

3. **In the Notes area, type any information you like.**

   You can enter contact information, notes about resources, or other useful information about the task.

4. **Format the note, if desired.**

   Click the buttons at the top of the Notes area to change the font. You can also

   - Left-align, center, or right-align text.
   - Format text as a bulleted list.
   - Insert an object.

5. **Click the OK button to save the note.**

   You will notice that a note icon appears in the information column indicating there is now a note there.

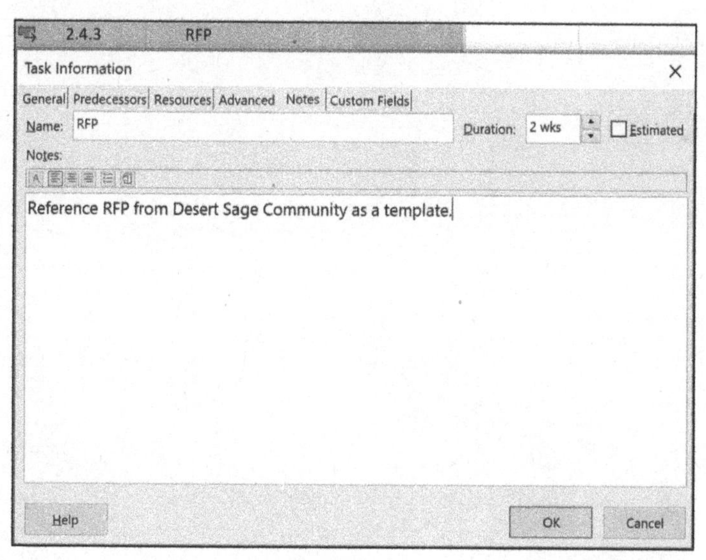


**FIGURE 3-8:** Creating a task note.

© *John Wiley & Sons, Inc.*

IN THIS CHAPTER

» **Identifying the different kinds of dependency relationships**

» **Creating dependency links**

» **Discovering how dependency links affect timing**

» **Allowing for lag and lead time**

» **Reflecting the timing of external tasks in your project**

# Chapter 4
# The Codependent Nature of Tasks

Imagine this: If you create a hundred tasks and leave their default constraints so that they start as soon as possible and have no dependencies, all those tasks start on the project start date and occur simultaneously. The project consisting of a hundred tasks takes exactly as long to complete as the longest task.

Wander back over here to reality for a moment. When did you ever have a set of tasks in which not a single task had to be completed before another one could start? Imagine what would happen if you tried to train your employees to use a new piece of equipment before the equipment even arrived.

The reality is that not all tasks in a project start at the same time. To reflect this reality in Project, you have to set dependency links between tasks. *Dependencies* are timing relationships between tasks — for example, when one task depends on the completion of another. Dependencies are caused by one of these factors:

» **The nature of the tasks:** You can't frame a house, for example, until its concrete foundation is dry; otherwise, the building will sink. This concept is sometimes referred to as a *mandatory dependency* or *hard logic*.

- **A best practice or preference:** You should document all requirements before you start designing, however — you can start *some* of the high-level design work before documenting every requirement; it just isn't a good practice to do so. This concept is sometimes referred to as a *discretionary dependency* or *soft logic.*
- **Resource availability:** When your operations manager can't attend two plant inspections simultaneously, the situation is known as a *resource dependency.*
- **Reliance on a resource or an action outside the project:** When you need a deliverable from another project before you can start your work, or you're waiting on someone from an external agency to sign off on it, you have an *external dependency.*

# How Tasks Become Dependent

In Chapter 2, I talk about manually scheduled tasks versus auto-scheduled tasks. If you allow the auto-scheduling and your network logic (dependencies) to build your schedule instead of manually assigning specific dates to tasks, Project can reflect changes to your schedule and adjust the dates and timing automatically.

For example, if the task of receiving materials is delayed by a week, the dependent task of starting the building process moves out a week automatically. You can note the change when you're tracking activity in your plan, and Project makes adjustments accordingly. The alternative is to change the start date of just about every task in your schedule *every time a task is running late*; you don't even want to think about doing that!

## Dependent tasks: Which comes first?

When we talk about task relationships, we talk about a task having a predecessor and successor. The predecessor task comes first and impacts when the *successor* (following) task is scheduled, particularly under the auto-scheduling method.

Figure 4-1 shows you how the taskbars in Gantt Chart view graphically depict the predecessors and successors in dependency relationships between tasks. Notice how taskbars represent the relationship when a task starts after another task. Also notice the lines drawn between tasks: These lines indicate *dependency links.*

Security system
Identify requirements
Alternatives analysis
RFP
Source Selection
Contract
System install
System test
System integration test
System complete
6/15

**FIGURE 4-1:** Dependencies.

© John Wiley & Sons, Inc.

Here's some important advice about dependencies: You can have more than one dependency link to a task, but don't overdo it. Many people who are new to Project make the mistake of building every logical timing relationship that can exist. If the situation changes and the dependencies have to be deleted or changed (for example, to shorten a schedule), the web of dependencies starts to get convoluted — and can easily create a nightmare.

For example, you must complete the tasks of obtaining a permit and pouring a foundation for a building before you can start framing it. However, if you set up a dependency between obtaining the permit and pouring the foundation, setting a dependency from foundation to framing is sufficient to establish the correct timing. Because you can't start pouring the foundation until you have a permit, and you can't frame until you pour the foundation, framing can't start before you have a permit. This common mistake is known as having a *redundant predecessor.*

You don't have to use dependencies to prevent resources from working on two tasks simultaneously. When you set the availability of resources and assign them to two tasks happening at the same time, you can use tools such as Team Planner view (see Chapter 9) and resource leveling (discussed in Chapter 12) rather than establish a dependency that forces one task to happen after another. Resource leveling delays tasks when scheduling causes a resource overallocation. See Chapters 9 and 12 for more about how resource assignments affect task timing.

# Dependency types

There are four types of dependency links: finish-to-start, start-to-start, finish-to-finish, and start-to-finish. Using these types efficiently can mean the difference between a project that finishes on time and one that's still limping along long after you retire.

Here's how the four dependency types work:

- **Finish-to-start:** This most common type of dependency link should account for about 90 percent of the dependencies you create. In this relationship, the predecessor task must be completed before the successor task can start. When you create a dependency, the default setting is finish-to-start.

  An example of a finish-to-start dependency is when you must complete the Source Selection task before you can begin the Contract task. All the dependencies shown in Figure 4-1 are finish-to-start. The relationship is indicated by a successor taskbar that starts where the predecessor taskbar ends. You see the finish-to-start type abbreviated as FS.

TIP

  The finish-to-start dependency is the default dependency. Therefore, you will not see FS (for finish-to-start) unless there is a lead or a lag (which are covered later in this chapter).

- **Start-to-start:** The start of one task is dependent on the start of another. In this dependency type, two tasks can start simultaneously, or one task might have to start before another task can start. For example, you can start purchasing the trucks and carts at the same time. A start-to-start relationship is abbreviated as SS.

  Figure 4-2 shows, accordingly, the start-to-start relationship between the Purchase Trucks and the Purchase Carts tasks. Note in the Predecessors column that Purchase Carts shows 19SS. That means the start is dependent on the start of Task 19, in this case Purchase Trucks.

- **Finish-to-finish:** In a finish-to-finish relationship, one task must finish before or at the same time as another task.

  Suppose you need to complete the staffing and the operations manual before you can conduct training. You might choose to show that the operations manual has to be done by the time the staffing is done. To do that you can indicate a finish-to-finish relationship between the Staffing and Develop Operations manual, as shown in Figure 4-3. This means if the finish of staffing moves, the finish of the operations manual will move. In the Predecessor column, you will see that finish-to-finish is abbreviated as 46FF (where 46 is the task ID for Staffing).

- **Start-to-finish:** In a start-to-finish dependency, the predecessor task can finish only after the successor task has started. If the successor is delayed, the predecessor task can't finish. Of course, this type of relationship is abbreviated as SF.

  Suppose that you're bringing online a new accounting application. The predecessor task of Turn Off Old Application can't be completed before the successor task of Start Up New Application has started. You need to make sure the new application works as expected, even though you have tested it — so you might run both for a month or so. Therefore, the demise of the old application is dependent on when the new application starts up.

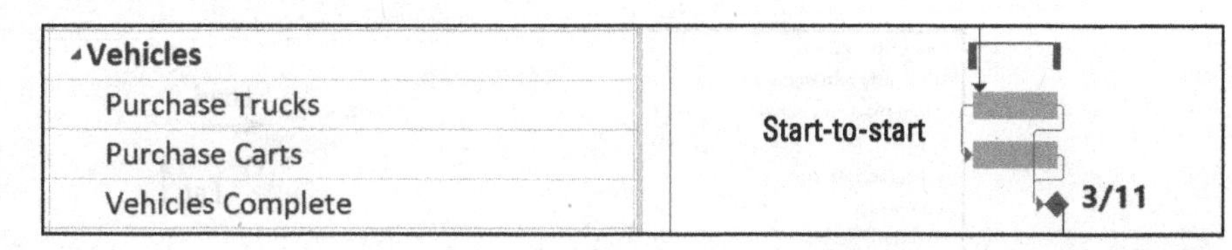


**FIGURE 4-2:** The start-to-start relationship.

© John Wiley & Sons, Inc.

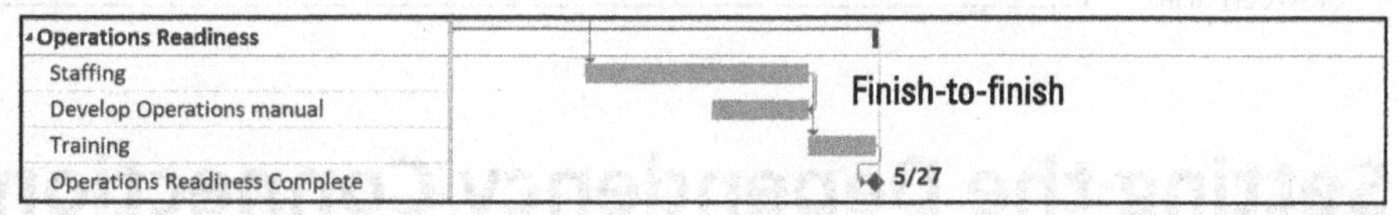


**FIGURE 4-3:** The finish-to-finish relationship.

© John Wiley & Sons, Inc.

WARNING

The start-to-finish dependency type is tricky. Leave this type of logic to the professional schedulers and try to find a different way to indicate the relationship between tasks. Often, simply breaking down the tasks and resequencing them accomplishes the same goal.

# Allowing for Murphy's Law: Lag and lead time

Dependencies can become a little more complex than simply applying the four types of dependency links that I describe in the preceding section. You can use lag time or lead time to fine-tune your timing relationships:

- **Lead time** is created when you accelerate time between the start or finish of a predecessor task. In Figure 4-4, the Request for Proposal (RFP) is set to start a week before the Alternatives Analysis task is done. This is a finish-to-start relationship with a one-week lead. Essentially, we are overlapping the RFP and the Alternatives Analysis tasks by a week. This is called *fast tracking*. You show a lead with a minus sign. So a finish-to-start relationship with a one-week lead would be shown as FS-1 wk (*wk* is the abbreviation for week).
- **Lag time** occurs when you add a delay the start or finish of a predecessor task; lag time causes a gap in timing, which delays the successor task. For example, a period almost always exists between sending out a Request for Proposal (RFP) for a service and receiving the proposals. Figure 4-4 shows the lag time scheduled between the RFP and Source Selection tasks. When you show a lag, you show the plus (+) sign and then the amount of lag. For example, a lag of three weeks is shown as +3wks. Therefore, the relationship between RFP and Source Selection is FS+3 wks.

**FIGURE 4-4:** Leads and lags between tasks.

© John Wiley & Sons, Inc.

# Setting the Dependency Connection

Setting dependency relationships is simple: You create a dependency, select the dependency type, and build in any lag time or lead time. The tricky part lies in understanding how each type of dependency affects your plan when your project goes live and you start to record actual work that resources perform on tasks.

## Adding the dependency link

When you create a dependency, it's a finish-to-start relationship by default: One task must finish before another can start. If that's just the kind of dependency you want, that's all there is to it. If it isn't the kind you want, after you create this link, you can edit it to change the dependency type or to build in lag or lead time.

To establish a simple finish-to-start link, follow these steps:

1. **Display Gantt Chart view and ensure that the two tasks you want to link are visible.**

   You may have to collapse some tasks in your project or use the Zoom button on the View tab to fit more tasks on the screen.

2. **Click the predecessor task and Ctrl+click the successor task. When both tasks are highlighted, click the Link Tasks button (see Figure 4-5) on the Task tab in the Schedule Group.**

   You can continue holding down the Ctrl key and highlighting as many tasks as you want to link. Figure 4-5 outlines the Link Tasks and Unlink Tasks buttons.

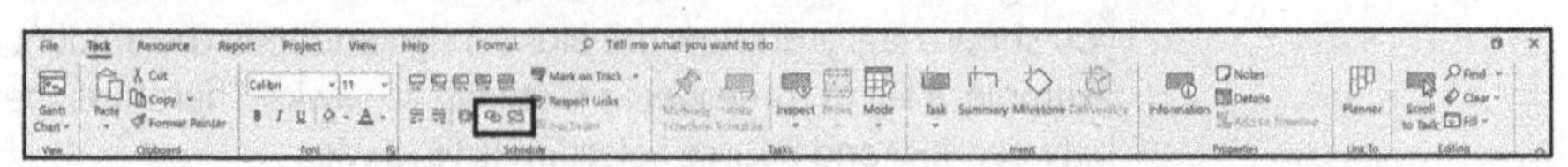

**FIGURE 4-5:** Linking and unlinking tasks.

© John Wiley & Sons, Inc.

You can link multiple tasks in a row in a finish-to-start relationship by clicking the first task and dragging to the last task. When you release the mouse button, click the Link Tasks button to link all the tasks in order.

WARNING

Best practice is to link only subtasks, which represent the actual work performed, not the summary tasks. The Ctrl+click method is helpful for skipping over summary tasks when selecting tasks for linking. Regardless of the method you use to select the tasks to link, the task that's selected first always becomes the predecessor. Therefore, if you're not careful in how you select the tasks, the dependency arrow could move backward in the schedule.

To establish a link in the Task Information dialog box or to modify an existing relationship, make note of the task ID number of the predecessor task and then follow these steps:

1. **Double-click the successor task.**

   The Task Information dialog box opens for the selected task.

2. **Click the Predecessors tab, as shown in Figure 4-6.**

   On this tab, you can build as many dependency relationships as you like.

3. **In the ID field, type the task ID number for the predecessor task.**

   In this case, we see that the predecessor of the Develop Operations manual task is Staffing.

4. **Press Tab.**

   The task name and the default finish-to-start dependency type showing 0d (no days, which is the default unit of time) of lag time are entered automatically.

5. **Click the Type column and click the arrow that appears to display the dependency types, and then click the appropriate dependency for your situation.**

   For this example, select Finish-to-Finish (FF).

6. **If you want to add lag or lead time, click the Lag field and use the spinner arrows that appear to set the amount of time.**

   Click up to a positive number for lag time, or click down to a negative number for lead time.

7. **Repeat Steps 3 through 6 to establish additional dependency relationships.**

8. **When you're finished, click the OK button to save the dependencies.**

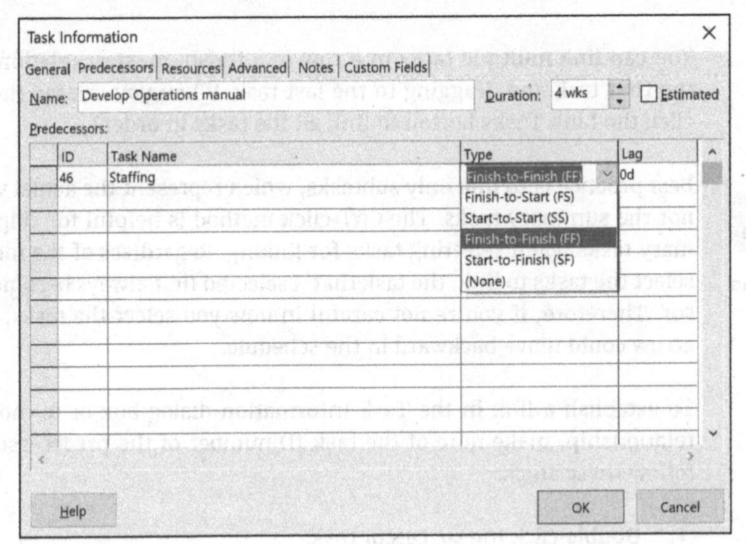


**FIGURE 4-6:** Setting dependency relationships.

© John Wiley & Sons, Inc.

**Here are some additional ways to set dependencies:**

- » Expand the Task Sheet pane until you see the Predecessors column. You can enter the task number of the predecessor task directly into that column, as shown in Figures 4-2, 4-3, and 4-4.
- » Alternatively, you can select the task from the Task Name drop-down list on the Predecessors tab of the Task Information dialog box. All the tasks you've already entered into the project appear. To display this list, click the next blank Task Name cell and then click the drop-down list arrow that appears, as shown in Figure 4-7.
- » To "keystroke" a link between tasks, select both tasks and then press Ctrl+F2. To unlink them, press Ctrl+Shift+F2.

# Words to the wise

**I can't stress enough how important it is to understand the nature of your dependencies. Assuming that you have identified the project work correctly, the way tasks are linked, coupled with your duration estimates (described in Chapter 5), is the key driver to developing an accurate schedule for your project. If you don't enter the correct relationship between tasks, or if you leave out a linkage, your schedule will be incorrect.**

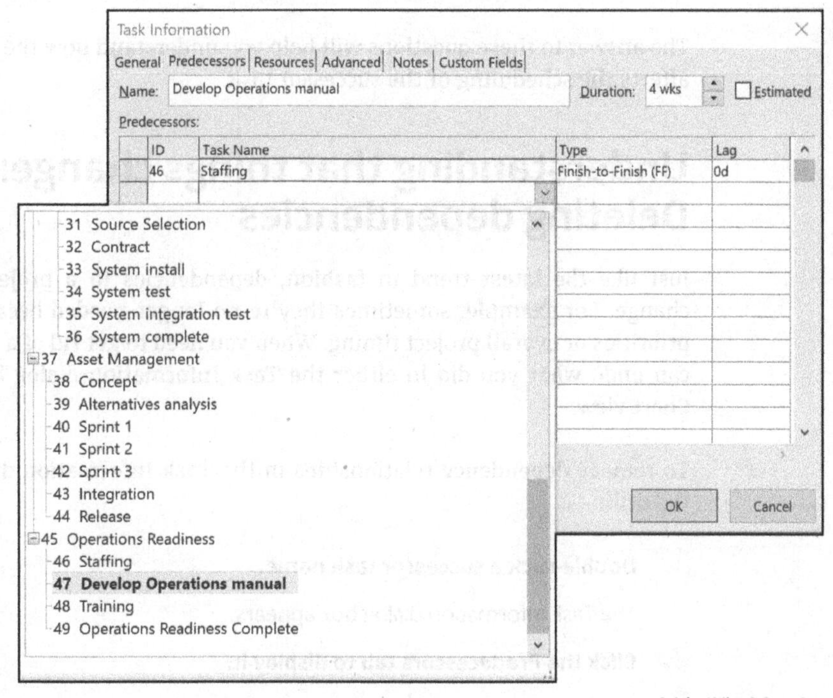


**FIGURE 4-7:** Setting dependency relationships with a drop-down list.

*© John Wiley & Sons, Inc.*

When all your tasks are logically linked, the result is a *network diagram*, or *precedence diagram* — a visual display of how all the tasks relate to one another. This diagram shows the flow of work through the project. If you're missing a relationship or dependency, the flow isn't right, and you're likely to be late or to have to scurry to try to make up for your oversight.

WARNING

Every task except the first and last tasks has a predecessor and a successor. If one of these elements is missing, the *hanger* task is left dangling. If a task has no predecessor, connect the task to the start milestone of the project. If the task has no successor, connect the task to the end of the phase or the finish milestone of the project.

In addition to making sure you understand the predecessor and successors of each task, you should make sure that the type of relationship is correct. If you try to save time by overlapping tasks, you should understand whether a FS relationship with a lead is a better option than an FF or SS with a lag. To help you determine the better option, ask yourself these questions:

- If the start of the predecessor is delayed, will it cause the successor to be delayed?
- If the finish of the predecessor is delayed, will it cause the successor to be delayed?

The answer to these questions will help you understand how the predecessor task affects the scheduling of the successor task.

## Understanding that things change: Deleting dependencies

Just like the latest trend in fashion, dependencies in a project can suddenly change. For example, sometimes they're no longer needed because of a shift in priorities or overall project timing. When you need to get rid of a dependency, you can undo what you did in either the Task Information dialog box or the Gantt Chart view.

To remove dependency relationships in the Task Information dialog box, here's the drill:

1. **Double-click a successor task name.**

   The Task Information dialog box appears.
2. **Click the Predecessors tab to display it.**
3. **Click the Type box for the dependency you want to delete.**

   A list of dependency types appears.
4. **Choose (None).**
5. **Click the OK button to save the change.**

   The dependency line on the Gantt chart is gone. The next time you open that Task Information dialog box, the dependency is gone, too.

TIP

Leading practices in project management suggest that dependencies should not be built based on resource availability when first creating the schedule. Rather, you should concern yourself with the logical order of the work regardless of who does it at this point in creating your schedule. *Resource leveling*, or the smoothing out of the peaks and valleys of resource usage, is always a factor in scheduling, of course, but it is a factor that will be addressed in later steps in project planning.

To remove dependency relationships with Gantt Chart view displayed, follow these steps:

1. **Select the two tasks whose dependency you want to delete.**
   - *For two adjacent tasks:* Drag to select their ID numbers.
   - *For nonadjacent tasks:* Click one task, press and hold Ctrl, and then click a nonadjacent task.
2. **Click the Unlink Tasks button in the Schedule group on the Task tab (refer to Figure 4-5).**

WARNING

Use the above method carefully. If you click only one task and then click the Unlink Tasks button, the result is somewhat drastic: *All* dependency relationships for that task are removed.

IN THIS CHAPTER

» Determining task types

» Effort-driven tasks

» Estimating effort and duration

» Using constraints

» Entering the start date

» Splitting tasks

Chapter 5

# Estimating Task Time

One of the most challenging aspects of managing a project is estimating how long a task will take to complete. Your team members usually make a good guess on task duration and hope for the best. But the estimate is where the concept of "garbage in, garbage out" (or, scarier, "garbage in, gospel out") truly applies. After all, the duration of individual tasks combined with network logic and the applied calendars determines the overall duration of the project — at least on paper.

An accurate estimate accommodates the nature of the task and employs the most applicable technique for estimating. Sometimes you estimate the effort involved, and sometimes you estimate the duration (the number of business days); it all depends on the nature of the task and the level of accuracy you need.

An accurate estimate isn't the only aspect of making a schedule realistic. Project helps your schedule reflect reality by allowing you to enter constraints, pause tasks, and split tasks.

REMEMBER

The schedule you build in Project is a model. It reflects what you think will happen — or what you plan for — given the information you enter. The more realistic the information you enter, the more realistic your schedule. However, ultimately, the schedule is a tool to help you and your team; it isn't reality. Do the best you can to acquire accurate data, but don't forget that you manage the schedule; it doesn't manage you.

## You're in It for the Duration

In projects, as in life, timing is everything. Timing in your projects starts with the durations that you assign to tasks. Although Project helps you see the effect of the timing of your tasks on the overall length of your plan, it can't tell you how much time each task will take. That's up to you.

Estimating the duration of tasks isn't always easy. Your estimate has to be based on your experience with similar tasks and your knowledge of the specifics of your project.

TIP

If your projects often share similar tasks, consider saving a copy of your schedule as a template that you can use in the future, thereby saving yourself the effort of re-estimating durations every time you start a similar project. Find out about saving templates in Chapter 21.

### Tasks come in all flavors: Identifying task types

Before you begin to enter task durations, be aware of task types in Project; they have an effect on how Project schedules the work of a task using the automatic scheduling mode after you begin assigning resources.

Essentially, your choice of task type determines which element of the task remains constant when you make changes to the task:

- **Fixed Units:** When you assign resources (Project considers them fixed units) to a task, they continue to work on that task even if the duration changes. (This type is the default.) For example, if you assign someone full time to build a presentation for senior management and you set the duration for three days, and then you realize that five days is more realistic, the number of fixed units stays the same. In other words, one person is still working on the task full time, but for five days rather than three days.
- **Fixed Duration:** The task takes a set amount of time to complete, no matter how many resources you add to the mix. For example, a test on a substance that requires leaving the test running for 24 hours has a fixed duration, even if you add 20 scientists to oversee the test.

» **Fixed Work:** The number of resource hours assigned to the task determines its length. If you set the duration of a Fixed Work task at 40 hours, for example, and you assign two resources to work 20 hours each (simultaneously) at units of 100 percent, the task will be completed in 20 hours. If you remove one resource, the single resource must spend 40 hours at units of 100 percent to complete the task.

You can see how understanding the various task types (along with how each one causes the task timing or resource assignments to fluctuate) is an important part of creating an efficient project.

Follow these steps to set the task type:

1. **Double-click a task.**

   The Task Information dialog box appears.

2. **Click the Advanced tab, as shown in Figure 5-1.**
3. **Choose one of the three options from the Task Type list.**
4. **Click the OK button.**

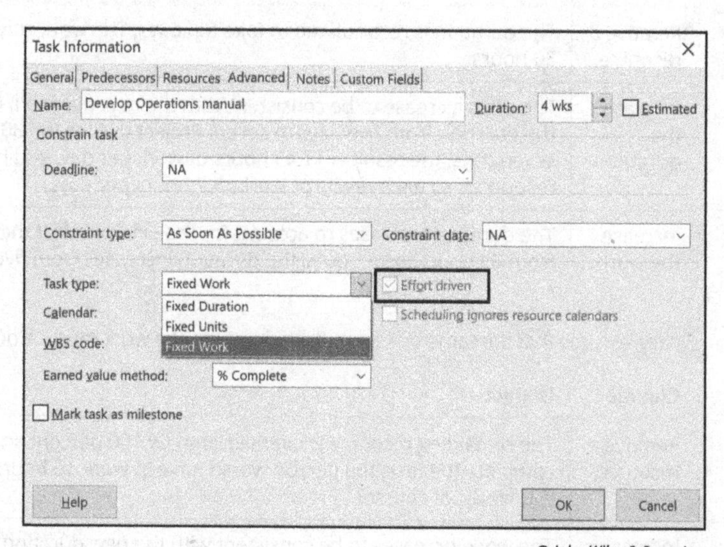


**FIGURE 5-1:** Setting the task type.

© John Wiley & Sons, Inc.

TIP

You can also display the Type column on a Gantt Chart sheet and set the task type there. To do this, right-click the top of any column, and select Insert Column from the contextual menu. A whole list of columns is available. Simply scroll down and select Type, and the column will be inserted.

## A TASK TYPE EXAMPLE

Because the task type is a complex concept, the example in this sidebar starts with a fixed-unit task that has a duration of 80 hours and two people assigned to the task full time for five days.

| Change | Impact |
|---|---|
| Remove a resource. | The duration is recalculated to ten days. The work stays the same at 80 hours. |
| Increase the duration. | The work increases to be consistent with the new duration. If you increase the duration from five days to seven days, you add 32 hours of work (two units × two 8-hour days). |
| Increase the work. | The duration increases to accommodate the extra work. If the work changes from 80 hours to 96 hours, for example, the duration increases from five days to six. |

Assume that the task type for the same task is Fixed Work. Thus, the work is set at 80 hours — no more, no less.

| Change | Impact |
|---|---|
| Remove a resource. | The duration is recalculated to take ten days. The work stays the same at 80 hours. |
| Increase the duration. | The units increase to be consistent with the new duration. If you increase the duration from five days to seven, Project divides the 80 hours by seven days, the result is 11.43 hours of work per day. You have two resources, so each resource works 5.7 hours per day. |
| Increase the work. | The duration increases to account for the extra work. If the work changes from 80 hours to 96 hours, the duration increases from five days to six. |

Finally, assume that the same task has a fixed duration: The work must be done in five days.

| Change | Impact |
|---|---|
| Remove a resource. | The remaining resource is overallocated by 100 percent. In other words, to complete the task, the person would have to work 16 hours per day (which is unlikely, of course). |
| Increase the duration. | The work increases to be consistent with the new duration. If you increase the duration from five days to seven, you add 32 hours of work (two units × two 8-hour days). |
| Increase the work. | The resources are either overallocated or you need to acquire an additional resource. If you have five days to accomplish 80 hours of work and you increase the work to 96 hours, your resources are 20 percent overallocated (96 ÷ 80 = 1.2). |

## Effort-driven tasks: 1 + 1 = ½

When you hear the word *effort* in Project, think *work*. For a task that's automatically scheduled, you can also set it up to be *effort driven*: If you adjust resource assignments, the duration might change, but the number of hours of effort (work) resources needed to complete the task stays the same. (Effort-driven scheduling isn't available for manually scheduled tasks.) When you add or delete a resource assignment on an effort-driven task, work is distributed equally among resources. (Refer to the description of fixed-work tasks in the preceding section.) In fact, all fixed-work tasks are effort driven. The more people you add, or the more hours they work, the shorter the duration, and vice versa.

Suppose that you need to set up a computer network in a new office in two days. You assign one resource who works 8 hours per day, so the work will take 16 hours to complete (two 8-hour days). If you then assign a second resource, this effort-driven task no longer takes two days, because the hours of effort required will be completed more quickly by the two people working simultaneously — in this case, in one 8-hour period.

Behind the scenes, effort-driven scheduling uses this formula to work this "magic":

Duration = Work ÷ Units

After you make the first assignment, any time you add or remove more units (people), Project recalculates the duration accordingly.

Select or clear the Effort Driven check box on the Advanced tab in the Task Information dialog box to enable or disable the Effort Driven setting; it isn't selected by default (refer to Figure 5-1). When you clear this check box, the same task that you set to run two days takes two days, no matter how much effort your resources contribute.

WARNING

Table 5-1 describes limiting behaviors in working with Project on effort-driven tasks.

**TABLE 5-1 Limiting Behaviors**

| Behavior | What Happens |
|---|---|
| First Assignment | When you first enter resources for an effort-driven task, the duration remains the same. If you add or delete resources after the task has been fully entered, the duration changes. |
| Fixed Work | Assigning additional resources reduces the task duration; reducing resources increases the duration. Fixed work is essentially effort-driven work, so you can't deselect the Effort Driven check box in fixed-work tasks. |
| Fixed Units | Assigning additional resources reduces the task duration; reducing resources increases the duration. |
| Fixed Duration | Assigning additional resources decreases the unit value of each resource; reducing resources increases the unit value of each resource. |
| Summary Tasks | Summary tasks can't be set to Effort Driven. |

# Estimating Effort and Duration

Developing accurate estimates — whether for resources, durations, or costs — is one of the most challenging and contentious parts of managing a project. You should understand the nature of estimating and the difference between the effort needed to accomplish the work and the duration, which indicates the number of required work periods (task duration). Several techniques are available to help you develop estimates, depending on the nature of the work. You can start by looking at the difference between effort and duration and then at the skills you need to develop accurate estimates.

*Effort* is number of labor units required to complete a task. Effort is usually expressed as staff hours, staff days, or staff weeks.

*Duration* is the total number of work periods (not including holidays or other non-working periods) required to complete a task. Duration is usually expressed as workdays or workweeks.

## Estimating techniques

Sometimes, people seem to estimate durations by snatching them out of the air or consulting a Magic 8 Ball. Estimating is undoubtedly an art and a science. The art stems from the expert judgment that team members and estimators bring to the process. Their experience and wisdom from past projects are invaluable in developing estimates, determining the best estimating method, and evaluating estimates (or the assumptions behind them) to assess their validity. In addition to the artful contribution of experts, team members, and estimators, a number of methods comprise the science of estimating. I discuss three of the more common methods used for Waterfall Projects in this section.

## Analogous estimating

*Analogous estimating* is the most common method of estimating. The aforementioned experts normally conduct this form of estimating. In its most basic form, this method compares past projects with the current project, determines their areas of similarity and areas of difference, and then develops an estimate accordingly.

A more robust application determines the duration drivers and analyzes the relationship between past similar projects with the current project. Duration drivers can include size, complexity, risk, number of resources, quality, or whatever other aspects of the project influence duration.

TIP

If you want to use analogous estimating effectively, your projects must be similar in fact, not simply similar in appearance. A software upgrade may sound similar to someone who is not familiar with software, but there are vast differences in what a software upgrade entails, so one software upgrade is not necessarily similar to others. The difference between moving from Windows 7 to Windows 10, for example, is much different from moving from Project 2019 to this version of Project; the magnitude of the work and, thus, the time is different!

## Parametric estimating

*Parametric estimating* uses a mathematical model to determine project duration. Though not all work can be estimated using this method, it's quick and simple: Multiply the quantity of work by the number of hours required to accomplish it. For example, if a painter can paint 100 square feet per hour and you have 6,000 square feet to paint, you can assume 60 hours of effort. If three people are painting (60 ÷ 3), the task should take 20 hours, or the equivalent of 2.5 days.

## Three-point estimating

When a lot of uncertainty, risk, or unknown factors surround an activity or a work package, you can use *three-point estimating* to produce a range and an expected duration. In this method, you collect three estimates based on these types of scenarios:

- **Best case:** In this optimistic (represented by the letter *O*) scenario, all required resources are available, nothing goes wrong, and everything works correctly the first time. You might see this type represented instead as $\mathbf{t_o}$ (for *time optimistic*).

» **Most likely:** The realities of project life are factored into the estimate, such as the extended unavailability of a resource, a work interruption, or an error that causes a delay. The *most-likely* (or M) scenario can also be represented by $t_m$ (time *most* likely).

» **Worst case:** This *pessimistic* (P) estimate assumes unskilled resources, or insufficient resources, a great deal of rework, and delays. It's represented by $t_p$ (for time *pessimistic*).

The simplest way to develop the expected duration, or $t_e$ (time *expected*) is to sum the three estimates and divide by 3. However, this technique isn't the most accurate one because it assumes — unrealistically — an equal probability that the best-case, most-likely, and worst-case scenarios will occur. In reality, the most-likely estimate has a greater chance of occurring than either the best-case or worst-case scenarios. Therefore, weight the most-likely scenario and determine the weighted average.

The most common way to calculate the weighted average is:

$$t_e = (t_o + 4t_m + t_p) \div 6$$

# Setting the task duration

Most tasks in a project (except milestones) have a duration, whether it's ten minutes or a year or another span of time. The needs of your project and the degree of control you require determine how finely you break down your tasks.

WARNING

If your project is to launch a satellite, for example, tracking the task duration by minutes on launch day makes sense. In most other circumstances, tracking the duration by days (or weeks, sometimes) is sufficient.

As with all task information, you can enter the duration in a Gantt Chart sheet or in the Task Information dialog box. Follow these steps to enter the duration in the dialog box:

1. **Double-click a task to display the Task Information dialog box.**
2. **If necessary, click the General tab to display it.**
3. **In the Duration box, use the spinner arrows to increase or decrease the duration, as shown in Figure 5-2.**
4. **If the current duration units aren't appropriate (for example, days when you want hours), type a new duration in the Duration box.**

A new task is created with an estimated duration of one day unless you change the duration. You can use these abbreviations for various time units:

- *m:* Minute
- *h:* Hour
- *d:* Day
- *w:* Week
- *mo:* Month

WARNING

Don't assume that changing the start and finish dates of a task changes its duration; it doesn't. You have to manually change the duration. If you don't, your project plan won't be what you intended.

5. **Click the OK button to accept the duration setting.**

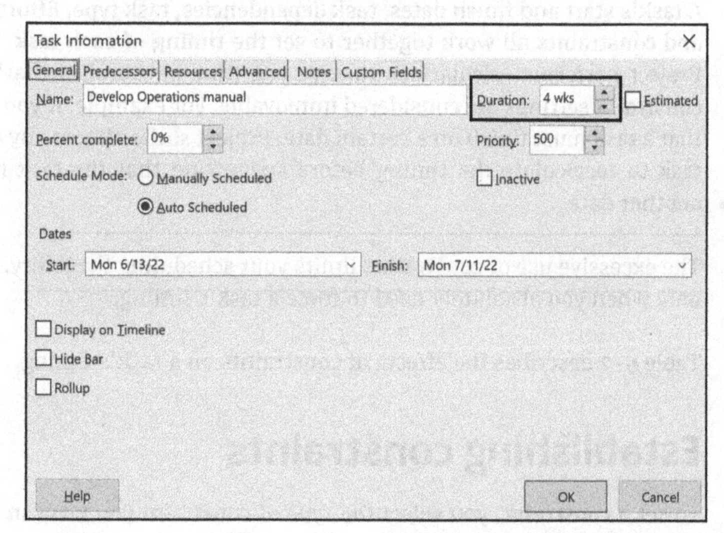


**FIGURE 5-2:** Setting the duration.

TIP

If you're unsure about the timing of a particular task, select the Estimated check box (on the General tab) when you enter the duration. This strategy alerts people to your lack of certainty. You can always go back and clear the box when you have better information.

## Controlling Timing with Constraints

In Project, *constraints* are timing conditions that control an automatically scheduled task. You tell Project which (if any) constraints are required for each task. Because manually scheduled tasks must follow the start and finish dates you specify for them, they already have de facto constraints.

### Understanding how constraints work

When you create a task and set it to be automatically scheduled, the As Soon As Possible constraint is selected by default. In other words, the task starts as soon as the project starts, assuming that no dependencies with other tasks exist that would delay it.

A task's start and finish dates, task dependencies, task type, Effort Driven setting, and constraints all work together to set the timing of each task. However, when Project performs calculations to save you time in a project that's running late, constraint settings are considered immovable. For example, if you set a constraint that a task must finish on a certain date, Project shifts almost any other scheduled task to recalculate the timing before suggesting that the task might finish on another date.

WARNING

The excessive use of constraints limits your scheduling flexibility. Use constraints only when you absolutely need to force a task's timing.

Table 5-2 describes the effects of constraints on a task's timing.

### Establishing constraints

To set a constraint, you select the type of constraint you want in the Task Information dialog box. Though you can set only one constraint for a task, some constraints work together with a date you choose. For example, if you want a task to start no later than a certain date, you select a date by which the task must start. Other settings, such as As Soon As Possible, work from a different date — in this case, the start date you set for the whole project or any dependency relationships you set up with other tasks. (See Chapter 4 for more about dependency relationships.)

**TABLE 5-2** **Task Constraints**

| Constraint | What Happens When You Apply It |
|---|---|
| As Soon As Possible | The task starts as early in the schedule as possible based on dependencies, calendars, and the project start date. (It's the default setting.) |
| As Late As Possible | The task occurs as late as possible in the schedule, based on dependencies, calendars, and the project's finish date. |
| Finish No Earlier Than | The end of the task can occur no earlier than the date you specify. |
| Finish No Later Than | The end of the task can occur no later than the date you specify. |
| Must Start On | The task must start on an absolute date. |
| Must Finish On | The task must finish on an absolute date. |
| Start No Earlier Than | The task can start no earlier than the date you specify. |
| Start No Later Than | The task can start no later than the date you specify. |

To set a task constraint, follow these steps:

1. **Double-click a task.**

   The Task Information dialog box appears.

2. **Click the Advanced tab.**
3. **Select a constraint from the Constraint Type list (see Figure 5-3).**
4. **If the constraint requires a date, select one from the Constraint Date list.**
5. **Click the OK button to save the settings.**

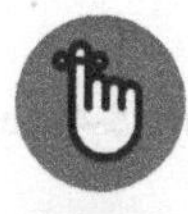

REMEMBER

The default constraint, As Soon As Possible, is already set, so you usually don't need to modify it.

TIP

When Project determines timing, a Must Start On constraint overrides the start date that is calculated based on start dates and durations.

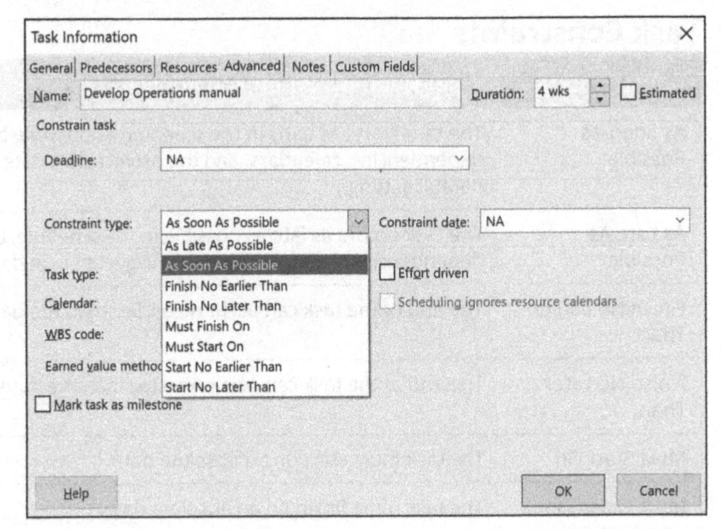


**FIGURE 5-3:** Setting a constraint.

© John Wiley & Sons, Inc.

## Setting a deadline

Sometimes I think that deadlines were made to be overlooked. Project agrees because, strictly speaking, deadlines aren't constraints (though the setting for the deadline is in the Constrain Task area of the Task Information dialog box, on the Advanced tab). Deadlines, which aren't the same as constraints, don't force the timing of task schedules. If you set a deadline and the task exceeds it, Project simply displays a symbol in the Indicator column to alert you so that you can panic — I mean, *take action* — appropriately.

To set a deadline, follow these steps:

1. **Double-click a task.**

   The Task Information dialog box appears.

2. **Click the Advanced tab.**
3. **Click the drop-down arrow in the Deadline field (shown in Figure 5-4) to display a calendar and then select a date.**

   If necessary, click the forward- or backward-facing arrow to move to a different month.

4. **Click the OK button to save the deadline setting.**

TIP

Display a Deadline column in the Gantt Chart sheet to enter the deadline or to show yourself and others your targeted deadline date.

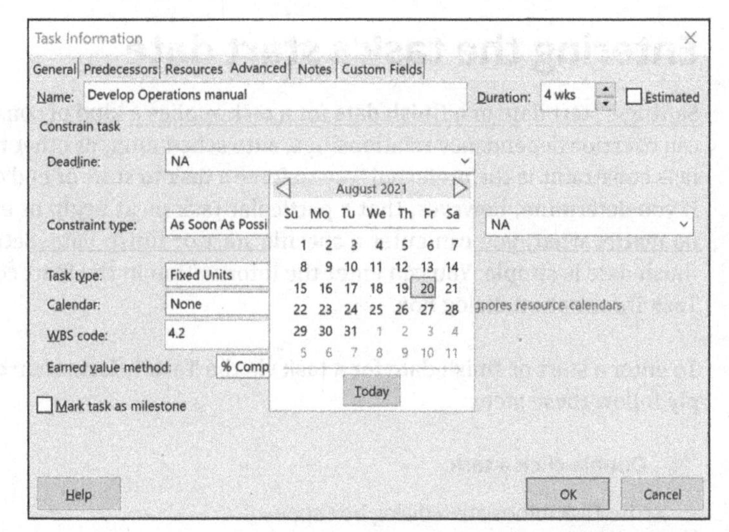


**FIGURE 5-4:** Setting a deadline.

© John Wiley & Sons, Inc.

# Starting and Pausing Tasks

When most people start using Project, they initially try to enter a start date for every task in the project. After all, you include dates when you create to-do lists, right? You're jumping the gun, though, and missing out on one of the great strengths of project management software: the capability to schedule tasks according to sometimes-complex combinations of factors, such as dependencies between tasks and task constraints. By allowing Project to determine the start date of a task, you allow it to make adjustments automatically when changes occur.

If you enter a task duration but not a start date for an automatically scheduled task, that task starts by default as soon as possible after the project start date that you specified in the Project Information dialog box, based on any dependencies you set up between tasks. For manually scheduled tasks, you eventually have to specify a start date to set the beginning schedule for the task.

Certain tasks, however, must start on a specific date. Examples are holidays, annual meetings, and the first day of the fishing season.

TIP

Project sets the finish date of a task based on when the task starts, the task duration, the task type, the resources assigned, and any calendars that have been set up (more about calendars in Chapter 8). If a task must finish on a certain date, however, you can set a finish date and let Project determine the start date.

## Entering the task's start date

Setting a start date or a finish date for a task applies a kind of constraint on it that can override dependency relationships, auto scheduling, or other timing factors. A task constraint is the preferred way to force a task to start or end on a certain day. If you determine, however, that a particular task must begin or end on a set date no matter what, you can enter a specific start or finish date. Setting the start or finish date is simple. You can enter the information in the Start column or via the Task Information dialog box.

To enter a start or finish date for a task via the Task Information dialog box, simply follow these steps:

1. **Double-click a task.**

   The Task Information dialog box appears.

2. **Click the General tab, if it's not already displayed.**
3. **Click the drop-down arrow at the end of the Start box or the Finish box.**

   A calendar appears, as shown in Figure 5-5.

4. **Click a date to select it, or click the forward- or backward-facing arrow to move to a different month and select a date.**

   If the current date is the date you want, take a shortcut and click the Today button on the drop-down calendar.

5. **Click the OK button.**

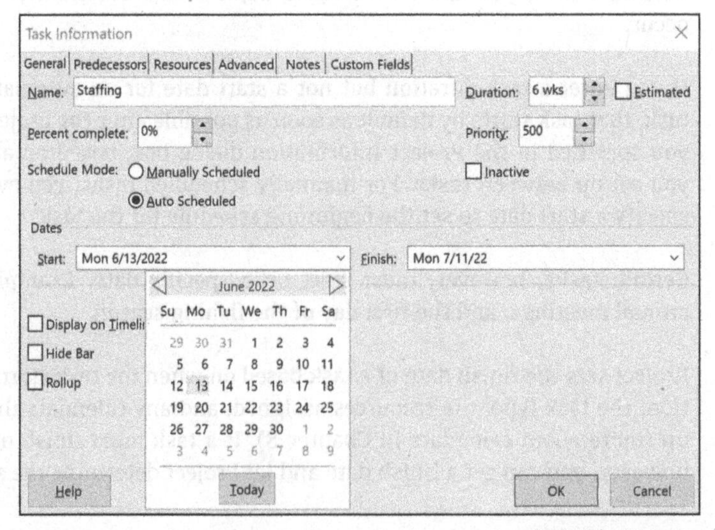


**FIGURE 5-5:** Setting a start or end date.

© John Wiley & Sons, Inc.

Be advised that when you are let Project auto-schedule your work and you set a start or finish date, a Planning Wizard dialog box will appear, as shown in Figure 5-6, letting you know that any task relationships will not drive the start of the task. You have three options:

- You can move the task to start on the date you set and remove the link.
- You can move the task to start on the date you set and keep the link.
- You can cancel this, not move the task, and keep the link.

If you are setting several dates manually, you may want to click the Don't Tell Me About This Again check box. Or, consider changing the task mode to Manually Schedule for those tasks that you are setting start or finish dates.

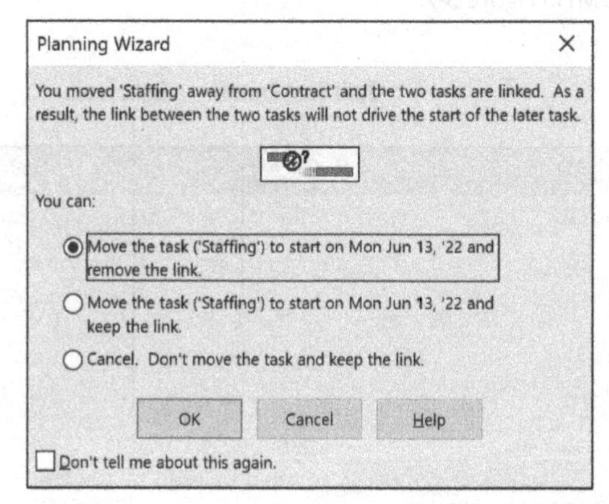


**FIGURE 5-6:** Planning Wizard box.

*© John Wiley & Sons, Inc.*

## Taking a break: Splitting tasks

Did you ever start a task — filing your taxes, for example — and find that you simply had to drop everything before you were done and go do something else? (In the case of filing my taxes, I usually need a break for a good cry.)

Projects work the same way. Sometimes, tasks start and then have to be placed on hold before they can start again later — for example, if you experience a work shutdown caused by labor negotiations. Or perhaps you can anticipate a delay in the course of a task and you want to structure it that way when you create it. In this case, you can use a Project feature to split a task so that a second or third portion starts at a later date, with no activity in the interim. You can place as many splits in a task as you like.

Follow these steps to split a task:

1. **On the Task tab of the Ribbon, click the Split Task button in the Schedule group.**

   A readout appears and guides you as you set the start date for the continuation of the task.

2. **Move the mouse pointer over the taskbar on the Gantt chart, and adjust the pointer's position until the box displays the date on which you want to start the task split; then drag to the right until the box contains the date on which you want the task to begin again.**

3. **Release the mouse button.**

   The split task shows up as a short task, a series of dots, and then the rest of the task, as shown in Figure 5-7.

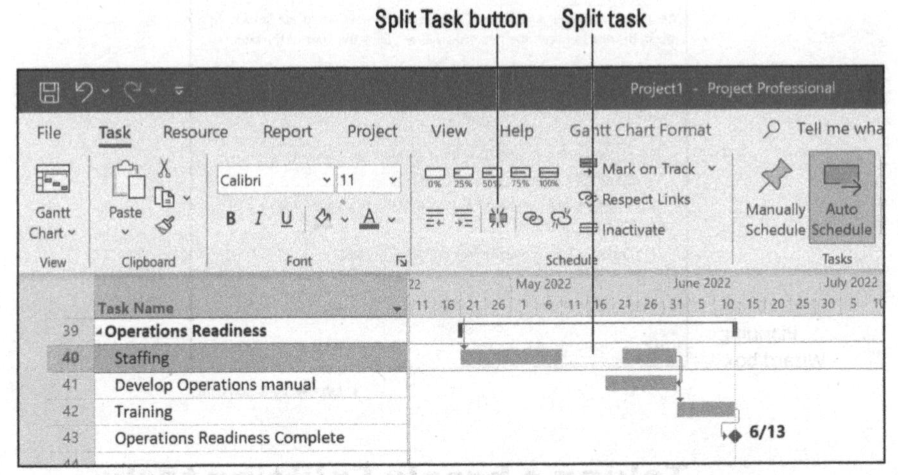


**FIGURE 5-7:** Splitting a task.

*© John Wiley & Sons, Inc.*

TIP

**To rejoin a split task, place the mouse cursor over the taskbar until the move cursor appears, and then drag the split taskbar backward to join with the other portion of the taskbar.**

REMEMBER

**Don't use the split-task approach to place an artificial hold on a task until another task is complete. Suppose that you start testing a product and then have to wait for approval before finalizing the test results. In this case, create a Testing task, a Final Approval milestone, and a Finalize Test Results task — and then create dependency relationships among them. This way, if a task runs late, your final task shifts along with it instead of being set in stone (as a split task can be).**

IN THIS CHAPTER

» Finding your way around Project

» Exploring Project views

» Understanding Network Diagram view

» Personalizing views for your needs

# Chapter 6

# Check Out This View!

In Project, a *view* is a way to display task and resource information. It's essentially a template with a predetermined set of columns in the worksheet, a way to enter data in a pane, or a way to view data in a chart. Though earlier chapters in this book describe how to enter task information into a worksheet in Gantt Chart view (the default view and the most common view), it is by no means the *only* way to look at data in Project. Project has lots of different views you can use to look at and analyze your project data.

After you know how to enter basic task information into Project, you can begin to display the information in different ways. First, I give you a high-level explanation of views and then dip into more detail about the more common views. Though I can't possibly cover every combination of data that you might want to see, I show you how to customize the views in Project to look at data in the way that best suits your needs at the time.

## A Project with a View

Views organize data so that you can see information in logical ways. Because of the complexity of information in a typical Project schedule, Project offers many views for slicing and dicing (organizing) the information. If an average word processing document is as complex as a cookie, your average Project file is more like a five-tier wedding cake adorned with intricate flowers and garlands in delicate swirls of sugary icing.

A Project file holds information about these elements:

- **Tasks:** Task name, duration, start and finish dates, assigned resources, costs, constraints, and dependencies, for example
- **Resources:** Resource name, resource type, rate per hour, overtime rate, department, cost per use, and others
- **Assignments:** Information about which resources will complete which tasks, including hours of work, units of effort, and remaining work
- **Timing and progress:** Several types of calendars, project start and finish dates, percentage of tasks completed, resource hours spent, baseline information, critical path information, and more
- **Financial information:** Earned value, time-and-cost variance, and projected costs for work not completed, for example

## Navigating tabs and views

For each project schedule file you create, Project essentially builds an extensive database of information. The different views in Project enable you to zero in on different combinations of the data you need. Having several views from which to observe your project information is helpful, but having all those views does you no good if you don't know how to move around in a view after you find it or how to move from one to another.

One way to move from one view to another is by using the Task tab on the Ribbon. Clicking the bottom section (with the down arrow) of the Gantt Chart button to the left of the Task tab displays a menu of Project's most-often-used views, as shown in Figure 6-1, as well as any custom views you create. Simply click any listed view to display it. In case it isn't obvious, clicking the top section of the Gantt Chart button redisplays Gantt Chart view.

You can see the same list of views by going to the Resource tab and clicking the View group's Team Planner drop-down menu.

Clicking the View tab on the Ribbon gives you all kinds of options for selecting a view. Using the Task Views group on the far left end, you can select the Gantt Chart, Task Usage, Task Board, Network Diagram, Calendar, or Other Views. The Resource Views group enables you to select the Team Planner, Resource Usage, Resource Sheet, or Other Views. Other groups on the View tab offer choices for sorting data, changing the timescale, and zooming a view to see it better. You can also split the view by adding the timescale to whichever view is active, or you can add details to a specific view.

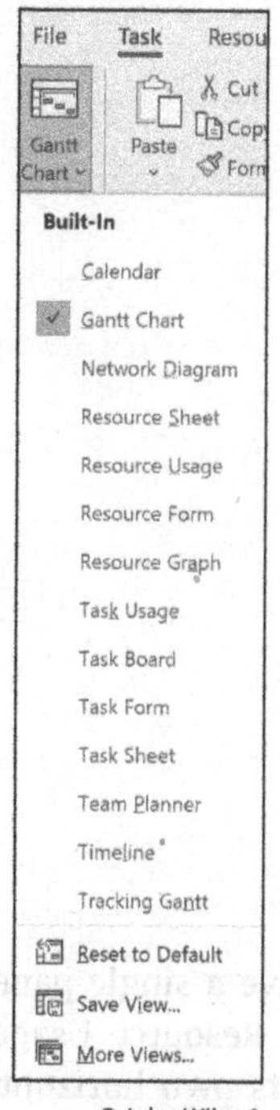


**FIGURE 6-1:** View choices.

© John Wiley & Sons, Inc.

TIP

The lower-right corner of the Project window has buttons for jumping directly to the Gantt Chart, Task Usage, Team Planner, Resource Sheet, and Report views.

In addition to the views you can display via the Task tab or the View tab, you may need to use one of the dozen or so other views as you work on your project. To display views not directly available via the Ribbon, follow these steps:

1. **Select either the Task tab or Resource tab on the Ribbon.**
2. **Click the down arrow on the Gantt Chart button.**
3. **Click More Views at the bottom of the menu.**

   The More Views dialog box appears, as shown in Figure 6-2.
4. **Use the scroll bar to locate the view you want.**
5. **Select the view you want and click the Apply button.**

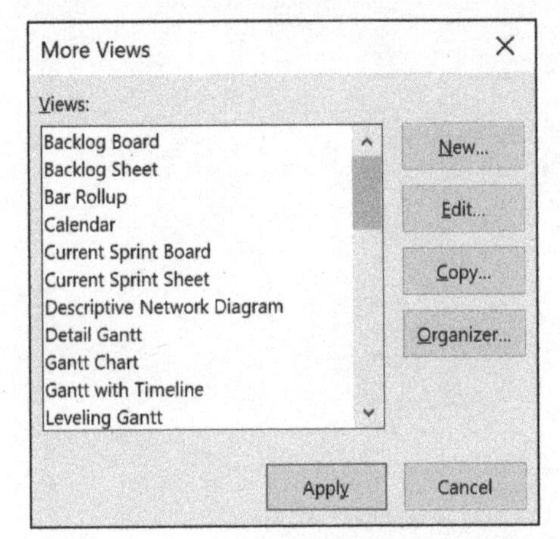


**FIGURE 6-2:** Available views in Project.

© John Wiley & Sons, Inc.

## Scrolling around

The simplest views, such as Calendar view, have a single pane, with horizontal and vertical scroll bars. Other views, such as Resource Usage view (shown in Figure 6-3), have two panes: Each pane has its own horizontal scroll bar, and because the panes share the vertical scroll bar, the panes move up and down together.

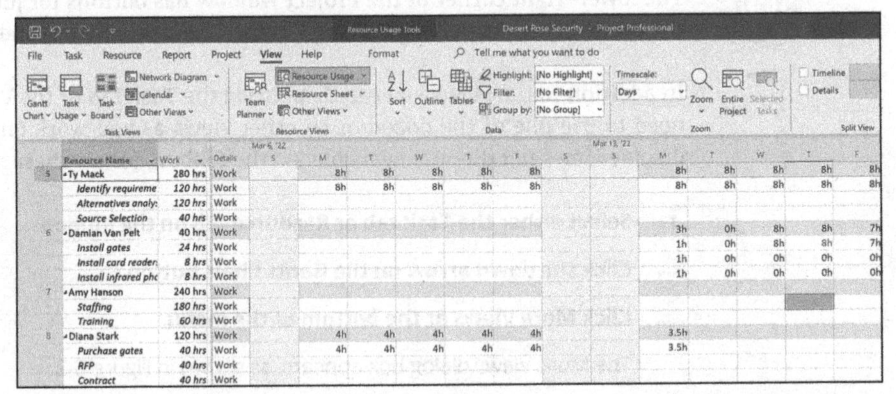


**FIGURE 6-3:** Multiple panes of information.

© John Wiley & Sons, Inc.

In most views with two panes, the pane to the left is the *sheet* — a spreadsheet-like interface that displays columns of information. To the right of this view is the *chart;* it uses bars, symbols, and lines to represent each task in your project and the dependency relationships among them.

At the top of the chart area is the *timescale.* This tool is used as a scale against which you can interpret the timing of the taskbars. To see your plan in greater or lesser timing detail, you modify the time units used in the timescale. For example, you can look at your tasks in detail over days or in a broader overview in months.

TIP

See the "Changing the timescale" section, later in this chapter, to find out how to change the time increments displayed in the timescale.

By using the horizontal scroll bars in each pane, you can view additional columns or additional periods in any pane with a timescale. Timescale panes cover the life of the project.

Use these methods to work with scroll bars:

- **Drag the scroll box on a scroll bar until you reach the location in the pane that you want to view.** When you drag the scroll box to move in a timescale display, the date display indicates where you are at any time in the scrolling calendar. Release the mouse button when the date display matches the date you want to view.
- **Click to the left or right of the horizontal scroll box to move one page at a time.** Note that a page in this instance is controlled to some extent by how you resize a given pane. For example, with a timescale pane and a timescale set to weeks, you move one week at a time. In a sheet pane that displays three columns, you move to the next (or preceding) column.
- **Click the right or left arrow at either end of a scroll bar to move in smaller increments.** On a sheet pane, depending on the size of your screen, you move about one half column per click.

# Reaching a specific spot in your plan

To reach a particular area of your Project schedule, you can also press the F5 key to use the Go To feature. Simply enter either of two items in the Go To dialog box when you want to find a task:

- A date from a drop-down calendar
- A task ID

You can also click the Scroll to Task button in the Editing group of the Task tab of the Ribbon (or press Ctrl+Shift+F5) to scroll the chart pane of the view. This shows the taskbar for a selected task in the sheet pane.

REMEMBER

The task ID, which is a number that's assigned automatically whenever you create a task, provides a unique identifier for every task in the plan.

TIP

To see more or less detail in your schedule, use the Zoom command on the View tab — in every view. Read more about using the Zoom button in the sidebar "Zooming in and out," later in this chapter.

For more about modifying the format of elements displayed in a view, see Chapter 13.

# More Detail about Views

Finding out how to use Project views to enter, edit, review, and analyze Project data can help you focus your attention on the aspects of the project you are interested in, whether it's resources, budget, task sequencing, or some other aspect. Don't worry that you'll be overwhelmed by the number of views you can use: After a while, using them will become second nature.

This section describes some common views: Gantt Chart, Resource Sheet, Team Planner, Network Diagram, Timeline, and Calendar.

## Home base: Gantt Chart view

Gantt Chart view is similar to a favorite room in your house — the place where your family members most often hang out. It's the view that appears first whenever you open a new project. Gantt Chart view, shown in Figure 6-4, is a combination of spreadsheet data and a chart with a graphical representation of tasks; it offers a wealth of information in one place. The spreadsheet can display any combination of columns of data that you want.

Gantt Chart view has two major sections: the sheet pane on the left and the chart pane on the right. In Gantt Chart view (or in any view with a sheet pane), you can use tables to specify which information is shown in the sheet. A *table* is a preset combination of columns (fields) of data that you can easily display by going to the View tab, and then clicking the Tables button in the Data group. From there you can choose one of nine preset tables:

- Cost
- Entry (the default table)
- Hyperlink

- Schedule
- Tracking
- Variance
- Work
- Summary
- Usage

You can also customize the column display of any table by displaying or hiding individual columns of data, one at a time. (See the section "Displaying different columns," later in this chapter, for information about this procedure.)

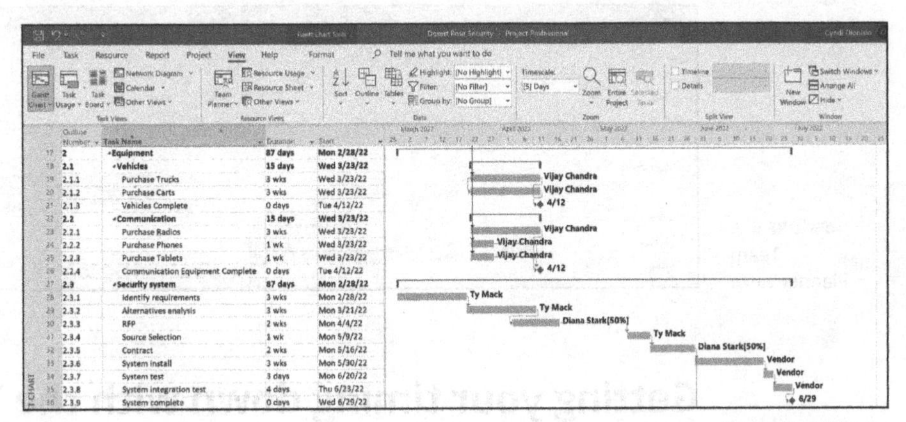


**FIGURE 6-4:** Gantt Chart view.

*© John Wiley & Sons, Inc.*

## Resourceful views: Resource Sheet and Team Planner

In Resource Sheet view, shown in Figure 6-5, you add the resources that will handle the work in your project. You can type entries in cells and press the Tab and arrow keys to move around. Chapter 7 describes the purpose of each field in this view and tells you how to set up your resource information correctly.

Team Planner view, shown in Figure 6-6, shows you what each team member is scheduled to work on, and when. You can change an assignment by simply dragging it from one resource to another. *Voilà* — assignment problems solved! You can find out more about this view in Chapter 9.

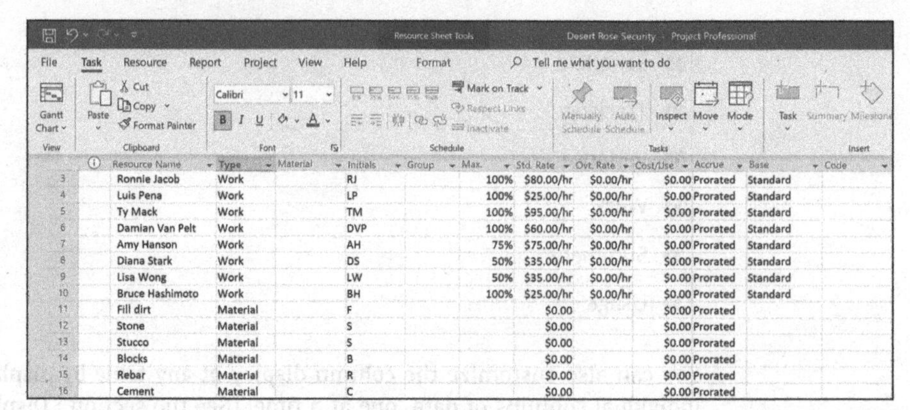


**FIGURE 6-5:** Resource Sheet view.

© John Wiley & Sons, Inc.

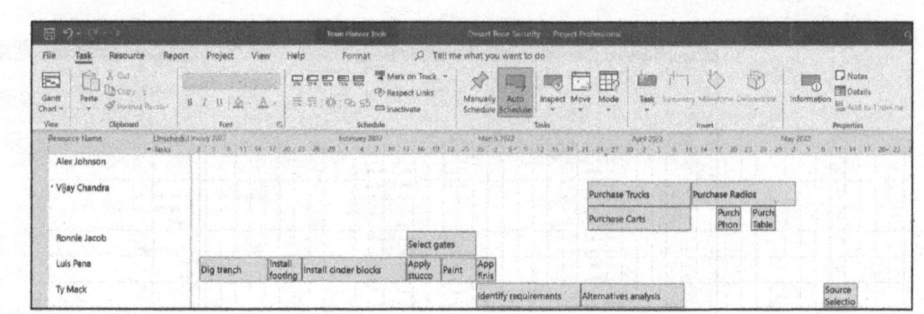


**FIGURE 6-6:** Team Planner view.

© John Wiley & Sons, Inc.

# Getting your timing down with the Timeline

Your sponsor and the project's client may not want to see every detail in your grand plan. If you need to share project information with a person whose eyes glaze over easily, meet your new best friend: the Project Timeline. Appearing in its own with certain other views, it presents a simplified picture of the entire schedule. You can also display individual tasks, parent tasks, or milestones on the Timeline. The Timeline in Figure 6-7 shows the summary-level tasks and key milestones. To display the Timeline, go to the View Tab, Split View group, and check the Timeline box. Chapter 18 delves into how to work with the Timeline and use it as a communication tool.

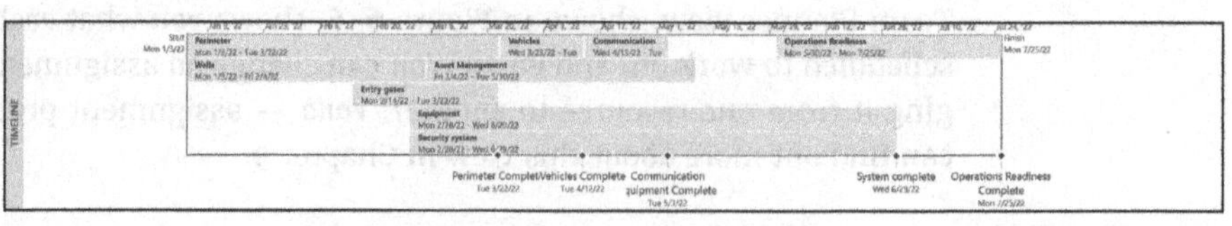

**FIGURE 6-7:** Timeline view.

© John Wiley & Sons, Inc.

## Going with the flow: Network Diagram view

The organization of information in Network Diagram view, as shown in Figure 6-8, represents the workflow in your project in a series of task boxes. The boxes include dependency lines that connect them to reflect the sequence of tasks. (See Chapter 4 for more about dependency relationships.) You read this view from left to right; earlier tasks on the left flow into later tasks and subtasks to the right. Tasks that happen in the same time frame are aligned vertically above each other. Tasks with an X through them have been marked as complete. Tasks with only one line through them have some progress but are less than 100 percent complete.

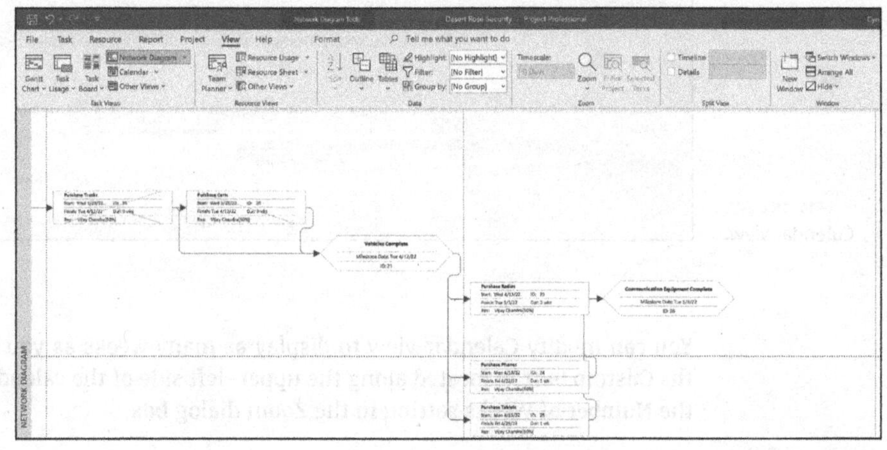


**FIGURE 6-8:** Network Diagram view.

*© John Wiley & Sons, Inc.*

**TIP**

Traditionally known as a *PERT chart*, the network diagram method was developed by the U.S. Navy in the 1950s for use in building the Polaris missile program. PERT stands for program evaluation and review technique.

Network Diagram view has no timescale; the view isn't used to see specific timing, but to see the general logical order of tasks in a plan. However, each task box, or *node*, holds specific timing information about each task, such as its start date, finish date, and duration. (You can customize the information in the task boxes, as described in the later section "Customizing Views.")

## Calling up Calendar view

Who can conceive of creating a schedule without opening a calendar? This familiar view of time is one of the many views offered in Project. Calendar view, as shown in Figure 6-9, looks like a monthly wall calendar, with boxes that represent days on a calendar in rows that represent the days in a week. Using this view, you can see all tasks that are scheduled to happen in any given day, week, or month.

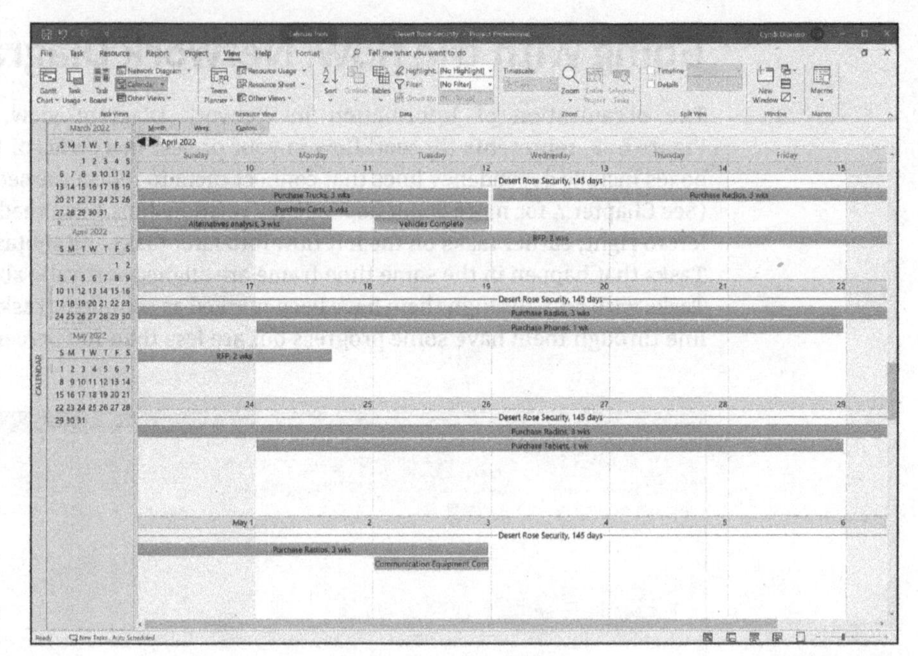


**FIGURE 6-9:** Calendar view.

© John Wiley & Sons, Inc.

You can modify Calendar view to display as many weeks as you need by clicking the Custom button located along the upper-left side of the calendar and changing the Number of Weeks setting in the Zoom dialog box.

REMEMBER

More than two dozen views are built into Project. You encounter many more as you work on specific elements of Project in this book.

# Customizing Views

Just when you thought you were starting to get a handle on the three dozen views available in Project, I dazzle you with even more possibilities: Every one of those views can be customized to show different types of information. For example, you can choose to show columns of information in spreadsheets, labels in Network Diagram boxes or taskbars, and sets of data in graph views. You can also modify the size of panes of information and adjust the timescale.

Why all this flexibility with the information you see onscreen? At various times in a project, you need to focus on different aspects of your tasks. If you're having a problem with costs, take a look at Resource Usage view and insert various columns of cost information, such as resource rates and total actual costs. If your plan is taking longer than the Hundred Years' War, consider displaying Tracking Gantt view and looking at a bunch of baseline columns with timing and dependency data or examining the project's critical path in the chart pane. If you need to display more of the sheet area so that you can read those columns without having to scroll, you can do that, too. In the following sections, you find out how to do all the things you need to do to show a variety of information in each view.

# Working with view panes

Like Gantt Chart view, several other views have two panes, such as Task Usage, Tracking Gantt, and Resource Usage. When the Timeline appears with a view, it appears in a third, horizontal pane at the top. You can modify the information you see in the sheet pane as well as the scale for timing in the chart pane. You can also display information near taskbars in the chart pane.

## Resizing a pane

In views that show more than one pane, you can reduce or enlarge each pane. This capability helps you see more information in one area, depending on your focus at the time.

REMEMBER

The overall area occupied by the panes is constant, so when you enlarge one pane, you reduce the other.

Follow these steps to change the size of a pane in a view:

1. **Place the mouse pointer over the edge of a pane.**
2. **When you see a pointer composed of two lines with arrows pointing in opposite directions, you can drag the pointer to resize the pane:**
   - *Drag to the left:* Enlarge the pane on the right.
   - *Drag to the right:* Enlarge the pane on the left.
   - *Drag up or down:* Change the size of the Timeline pane versus the area that's available for the rest of the view.
3. **Release the mouse button.**

   The panes are resized.

## Changing the timescale

I wish I could tell you that Project lets you control time and bestow lots more of it on your project, but it doesn't. What it does allow you to do is modify the timescale to display your plan in larger or smaller time increments.

A *timescale* consists of a possible total of three tiers that you can use to display different time increments. For example, the top tier can mark months; the middle tier, weeks; and the bottom tier, days. This variety of detail lets you easily observe the overall task length and points in time during the life of the task. You can use all three tiers, only the middle tier, or the middle and bottom tiers (as shown in Figure 6-10). This feature is especially helpful because projects can span several weeks to several years, such as construction or pharmaceutical projects. This scaling feature is the only way to view such large-scale efforts.

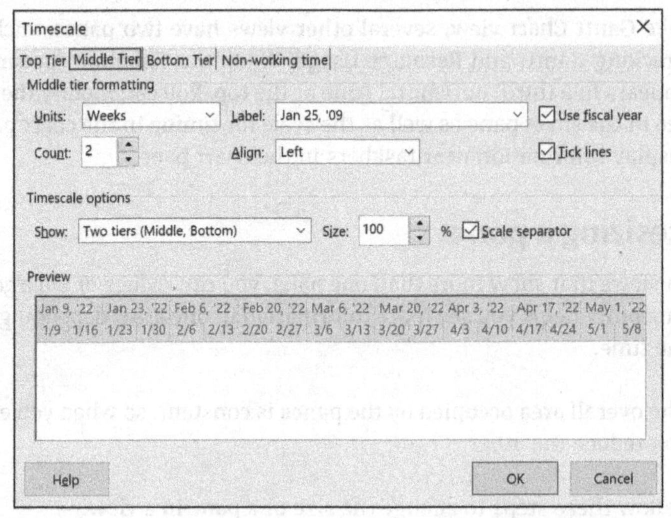


**FIGURE 6-10:** The Timescale dialog box.

*© John Wiley & Sons, Inc.*

You can modify the units of time and the alignment of each tier and also add tick lines to mark the beginning of each increment on the timescale. You also decide whether to include nonworking time on the timescale. For example, if you add an indication of nonworking time on a project for which weekends are nonworking, Saturdays and Sundays are indicated by a shaded area in the display, which can make a useful visual divider between weeks.

Here's how to access the Timescale dialog box and modify the timescale:

1. **Click the View tab.**
2. **Click the Timescale at the bottom of the Timescale drop-down menu in the Zoom group.**

   The Timescale dialog box opens (refer to Figure 6-10).
3. **Click a tier tab and select a style for Units, Label, and Align.**

4. **Set the count.**

   For example, if your Units choice is weeks and you change the Count option to 2, the timescale appears in two-week increments.

5. **To prevent the display of a certain tier, select One Tier or Two Tiers from the Show drop-down list in the Timescale Options section.**
6. **If you want Project to use fiscal-year notation in the timescale, select the Use Fiscal Year check box.**

TIP

Set the start date of the fiscal year by choosing File⇨Options⇨Schedule.

7. **To show marks at the beginning of each unit of time, select the Tick Lines check box.**
8. **Repeat Steps 2 through 6 for each tier that you want to modify.**
9. **Click the Non-Working Time tab.**
10. **In the Formatting options area, select the option you prefer.**

    Your choices are to have the shaded area for nonworking time appear behind taskbars or in front of taskbars or to simply not appear.

11. **In the Color list or Pattern list, select different options for the shading format.**
12. **Click the Calendar setting and select a different calendar on which to base the timescale.**

    You can find out more about calendar choices in Chapter 8.

13. **Click the OK button to save your new settings, which apply only to the timescale for the currently displayed view.**

TIP

Use the Size setting on the three tier tabs to shrink the display proportionately and to squeeze more information on your screen or the printed page.

TIP

Use these keystroke combinations to navigate the timescale:

- **Alt+Home:** Move to the beginning of the project.
- **Alt+End:** Move to the end of the project.
- **Alt+right arrow:** Scroll the timescale to the right.
- **Alt+left arrow:** Scroll the timescale to the left.
- **Alt+Page Down:** Move the timescale one page to the right.
- **Alt+Page Up:** Move the timescale one page to the left.

## ZOOMING IN AND OUT

One way to modify the display of any Project view is to use the Zoom menu in the Zoom group on the View tab of the Ribbon. This feature shows you a longer or shorter period in your project without your having to change the timescale settings manually. When you need to see several years at a time in your project, for example, click Zoom (the magnifying glass icon) and then zoom out several times until as many months or years as you like can fit into the view.

You can also click the Entire Project button in the Zoom group to show the entire project on the screen. Another nifty technique is to click the Zoom Selected Tasks button. The chart moves so that the bar representing the task is shown on the chart screen. Another quick (but less precise) way to zoom is to use the Zoom slider in the lower-right corner of the Project window. Drag it left to zoom out and drag right to zoom in.

## Displaying different columns

Each spreadsheet view has certain default columns of data that are stored in tables. Gantt Chart view with the Tracking table displayed, for example, has data related to the progress of tasks. The Resource sheet contains many columns of data about resources that can be useful for entering new resource information. In addition to choosing to display predefined tables of columns, you can modify any spreadsheet table to display any columns you like.

To insert a column, go to the Format tab. In the Columns group, click Insert Column. Another option is to follow this procedure to show selected columns of data:

1. **Scroll the table to the right and click the Add New Column heading.**

   A list of available columns appears. The available columns are shown in alphabetical order — I'm only showing you 34 here and I'm not even through the Bs! So as you can see in Figure 6-11, there are quite a few options to choose from!

2. **In the list, select the field that contains the information you want to include.**

   Typing the first letter of the field name scrolls the list alphabetically.

3. **If you want to enter a different label for the field heading, right-click the field heading and choose Field Settings; type the name in the Title box and click the OK button.**

   The title in the current view appears in the column heading for this field.

4. **Move the mouse pointer over the column heading, select the column, drag the column to the left or right, and reposition the column where you want it relocated within the table.**

[Type Column Name]
% Complete
% Work Complete
Active
Actual Cost
Actual Duration
Actual Finish
Actual Overtime Co
Actual Overtime Wo
Actual Start
Actual Work
ACWP
Assignment
Assignment Delay
Assignment Owner
Assignment Units
Baseline Budget Co
Baseline Budget Wo
Baseline Cost
Baseline Deliverable
Baseline Deliverable
Baseline Duration
Baseline Estimated
Baseline Estimated
Baseline Estimated
Baseline Finish
Baseline Fixed Cost
Baseline Fixed Cost
Baseline Start
Baseline Work
Baseline1 Budget C
Baseline1 Budget W
Baseline1 Cost
Baseline1 Deliverab
Baseline1 Deliverab
Baseline1 Duration
Baseline1 Estimatec
Baseline1 Estimatec
Baseline1 Estimatec
Baseline1 Finish
Baseline1 Fixed Cos
Baseline1 Fixed Cos
Baseline1 Start
Baseline1 Work
Baseline10 Budget
Baseline10 Budget
Baseline10 Cost
Baseline10 Delivera
Baseline10 Delivera
Baseline10 Duration
Baseline10 Estimate

**FIGURE 6-11:** New columns to insert.

© John Wiley & Sons, Inc.

TIP

**To hide a column, right-click its heading in the sheet pane and then choose Hide Column. Or, to insert a column at a particular location, right-click the column heading at that location and choose Insert Column. It's then inserted directly to the left of the selected column.**

TIP

**You can display preset tables of sheet data, such as Tracking for recording progress on tasks or Entry for entering new task information: Simply choose View⇨Tables (in the Data group) and then click the name of the table you want to display. The default table is the Entry table.**

## Modifying Network Diagram view

Network Diagram view is useful for understanding the flow of tasks throughout your project, and it can help you understand which series of tasks determines the overall project duration. Known as the *critical path*, it's the longest chain of tasks (in duration, not in number of tasks) in your project. The tasks on the critical path are *critical* tasks; tasks that aren't on the critical path are *noncritical* tasks.

When you first display Network Diagram view, you see boxes of various types — one for each task in your project. You can modify the information contained in those boxes and modify the field formatting.

By default, a typical task contains the task name, task ID, start date, finish date, duration, and resource name. A milestone displays only the milestone date, milestone name, and task ID number.

Different categories of tasks (such as *critical* or *noncritical*) can contain different information, and you can modify the field information contained in any individual box or category of boxes.

### Changing what's in the box

Sometimes, you want to see information about task timing, and at other times, you focus on other issues, such as resources. To accommodate these various information needs, Network Diagram view allows you to use various templates for the information contained in the diagram boxes.

To modify the information included in these boxes, follow these steps:

1. **Right-click anywhere in Network Diagram view outside any box and choose Box Styles in the right-click menu.**

   The Box Styles dialog box appears, as shown in Figure 6-12.

2. **In the Style Settings For list, select a task category.**
3. **To modify the data included in the task boxes, select a different template from the Data Template list.**

TIP

   You can choose additional templates (and edit any template to include whatever data you like) by clicking the More Templates button.

   A preview of the data included in the template appears in the Preview area.

4. **Click the OK button to save the new template.**

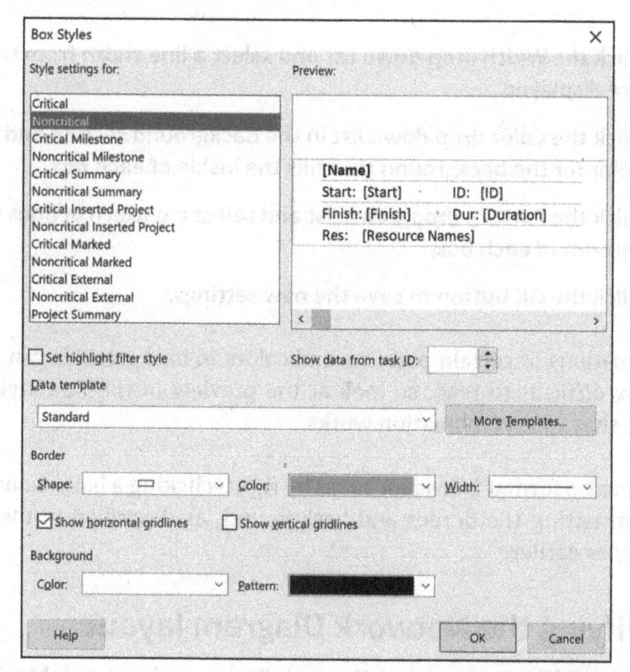


**FIGURE 6-12:** The Box Styles dialog box in Network Diagram view.

*© John Wiley & Sons, Inc.*

TIP

You can also apply different templates to each box separately by right-clicking a box (rather than clicking outside it), choosing Format Box, selecting a different template from the Data Template drop-down list, and then clicking the OK button.

## Beautifying diagram boxes

Are you the creative type? Do you dislike the shape or color of the boxes in Network Diagram view? Do you want to add shading to the box backgrounds? Project lets you do all this and more.

Follow these steps to modify the format of the Network Diagram boxes:

1. **Right-click anywhere in Network Diagram view outside any box and choose Box Styles in the right-click menu.**

   The Box Styles dialog box appears, as shown in Figure 6-12.

2. **Click the Shape drop-down list and select a different shape from the list that appears.**

3. **Click the Color drop-down list in the Border section and select a different color from the list of colors.**

   ***Note:*** This choice specifies the color of the line that forms the box, not a background color. For that topic, see Step 5 in this list.

4. **Click the Width drop-down list and select a line width from the ones that are displayed.**
5. **Click the Color drop-down list in the Background section and select a color for the background that fills the inside of each box.**
6. **Click the Pattern drop-down list and select a pattern of lines to fill the interior of each box.**
7. **Click the OK button to save the new settings.**

REMEMBER

Combinations of certain patterns and colors in backgrounds can make the text in the box difficult to read, so look at the preview in the Box Styles dialog box to ensure that your combination works.

TIP

You can also format *individual* boxes by right-clicking a box, choosing Format Box, and formatting the border and background, as described in the information on Box Styles earlier.

## Modifying the Network Diagram layout

You can modify the layout of Network Diagram view by right-clicking outside a box and selecting Layout. The Layout dialog box opens, as shown in Figure 6-13. You can either allow Project to position the boxes or position them manually. You can also define how to lay out the boxes and whether to show summary tasks.

Cosmetic options include defining the link style and whether to show link labels or connector arrows. You can set the critical path to show in a specific color (red is the default) and change the background color. A neat trick is to edit the layout to show link labels, such as FS (for finish-to-start). When you check the Show Link Labels check box, the type of dependency between tasks is shown.

You can also collapse the boxes so that only the task number appears. Simply right-click outside a box and select Collapse Boxes. You might also want to zoom the Network Diagram view to show as much information as you need. See the "Zooming in and out" sidebar earlier in this chapter. Figure 6-14 shows collapsed diagram boxes and the link labels on the dependency lines.

# Resetting the view

Most changes you make in a view apply as long as you have saved them. But if you've made so many changes that the view is now messy, you can reset it at any time: Click the Task tab on the Ribbon, click the bottom of the Gantt Chart button, and click Reset to Default. Once you click Yes, the view reverts to the default settings.

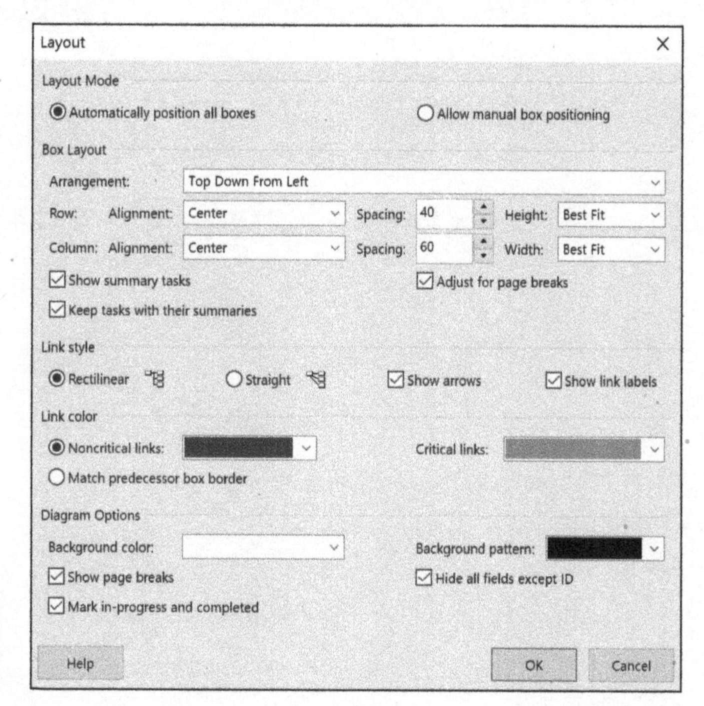


**FIGURE 6-13:** Modifying the Network Diagram layout.

© *John Wiley & Sons, Inc.*

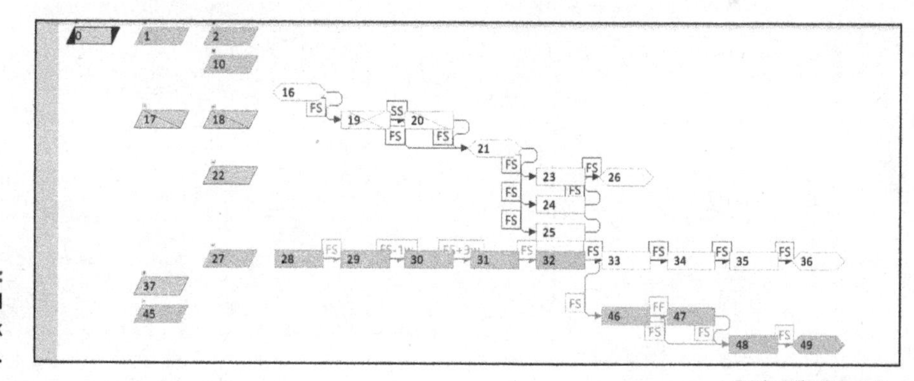


**FIGURE 6-14:** The modified Network Diagram view.

© *John Wiley & Sons, Inc.*

# 2 Managing Resources

IN THIS PART . . .

Identify the resources you need to complete the project.

Create and share resources.

Assign resources to get the work done.

Determine project costs.

IN THIS CHAPTER

» Understanding work and material resources

» Creating resources to complete your project

» Setting resource availability

» Benefitting from resource groups and shared resources

# Chapter 7
# Creating Resources

Projects utilize people, equipment, and materials, and these people, equipment, and materials are called project *resources*.

Using resources wisely is the delicate art of combining the right resources with the right skills and assigning those resources to invest the right amount of time (or units) for tasks. And you have to do this without overbooking anyone or anything at any point in the schedule.

A resource, because it affects timing and costs, is a big deal in Project. For this reason, many tools are available to help you create resources, specify how and when they'll work, assign them to tasks, manage their costs, and manipulate their workload. The first step in working with resources is to create them and enter certain information about them. That's what this chapter is all about.

## Resources: People, Places, and Things

Many people hear the word *resources* and think that it means *people*. Well, people are indeed frequently used as project resources, but they're only part of the story. Resources can also consist of equipment you rent or buy or materials such as wood or steel. You can even create resources that represent facilities you have to rent by the hour, such as laboratories or meeting spaces.

Here are some typical project resources (and one not-so-typical resource):

- Engineer
- Office supplies
- Hotel ballroom
- Administrative assistants
- Furniture
- Computer software
- Graphical designer
- Rocket fuel

You get the picture. Resources can be practically anything or anyone that you use to complete your project.

TIP

You can also create a resource as a cost placeholder if you're tracking costs using Project. For example, you might assign a resource named Plant Visit a cost of $1,750 to cover airfare, hotel, and rental car charges for the visit.

# Becoming Resource-Full

After you create and organize the tasks in your project, the next step, typically, is to create its resources. Before you create resources in Project willy-nilly, though, you must understand how they affect your project.

## Understanding resources

First and foremost, a *resource* is an asset that helps accomplish a project, whether the asset is a person, piece of equipment, material, or a supply. One aspect of working with resources effectively is to manage the workflow of any resource that has limited time availability for your project.

When you create resources, you indicate their availability by hours in the day or days of the week. For example, one person may be available 50 percent of the time, or 20 hours in the standard 40-hour workweek, whereas another may be available full time (40 hours). When you assign these kinds of resources to your project, you can use various views, reports, and tools to see whether any resource is overbooked at any point during the project. You can also see when people are sitting around, twiddling their thumbs, or when they may be available to help on another task. You can even account for resources that work on multiple projects across your organization and ensure that they're being used efficiently.

Another aspect of working with resources effectively is understanding how the number of resources you assign to work on a task affects the duration of that task. In other words, if you have a certain amount of work to perform but few people to do that work, a typical task takes longer to finish than if scads of folks are available.

REMEMBER

The task type determines whether a task's duration changes based on the number of resources assigned to it. The task type can be fixed units, fixed duration, or fixed work. (Chapter 5 describes task types in more detail.)

Finally, resources add costs to a project. To account for costs in your project — such as a person working many hours on a task, computers that you have to buy, or services that require fees — you must create resources and assign them to one or more tasks. I discuss project costs in Chapter 10.

## Resource types: Work, material, and cost

For the purposes of resource planning, Project recognizes only these three types of resources:

- **Work:** This person, such as a programmer or plumber, can be reassigned but not depleted. Work resources are assigned to tasks to accomplish work.
- **Material:** Material has a unit cost and doesn't consume working hours, but it can be depleted. Examples of material costs are steel, wood, and gravel.
- **Cost:** Using a cost resource gives you the flexibility to specify the applicable cost, such as travel or shipping, every time you use the resource.

TIP

Deciding which resource type to use when adding an external vendor or equipment resource to the project can be tricky. If you want to avoid adding hours of work to the project when that resource is being used, don't set up the resource as a work resource. Use a cost resource or fixed-cost approach instead.

## How resources affect task timing

Under Project's default settings, effort-driven scheduling is turned off for both manually and automatically scheduled tasks. So no matter how many work resources you assign, the task duration stays the same and Project piles on more hours of work for the task, to reflect more effort.

However, in some cases of a fixed-unit or fixed-work task type, the addition or removal of resources assigned to the task should have an impact on the time it takes to complete the task. In essence, the old maxim "Two heads are better than one" may be modified to "Two heads are faster than one."

Suppose that one person is assigned to the Dig Ditch task, which requires four hours of effort. Two people assigned to the Dig Ditch task will finish the job in two hours because two hours are being worked by each resource simultaneously, which achieves four hours of effort in half the time. To make scheduling a task work this way, you first have to change the task to an auto-scheduled task by clicking the Auto Schedule button in the Tasks group on the Task tab. Then you have to turn on effort-driven scheduling by double-clicking the task, clicking the Advanced tab in the Task Information dialog box, selecting the Effort Driven check box, as shown in Figure 7-1, and then clicking the OK button. (I discuss effort-driven tasks in Chapter 5.)

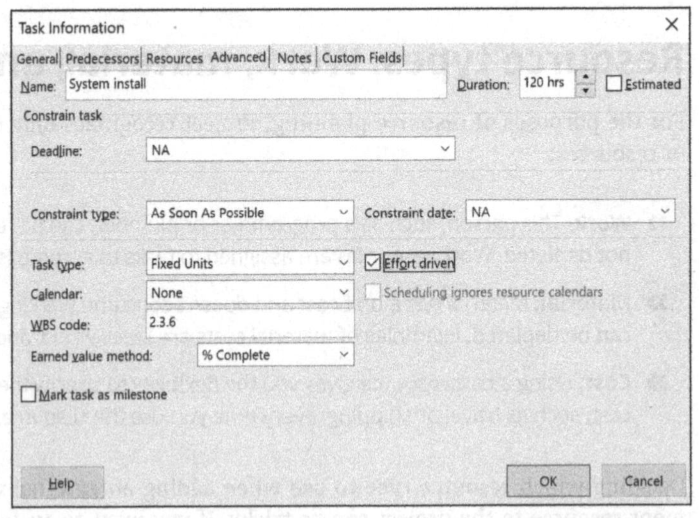


**FIGURE 7-1:** The Effort Driven check box.

WARNING

Assigning additional people to tasks doesn't always shorten work time proportionately, even though that's how Project calculates it. When you have more people, you also have more meetings, memos, duplicated effort, and conflicts, for example. If you add more resources to a task, consider also increasing the amount of effort required to complete that task to account for inevitable workgroup inefficiencies.

## Estimating resource requirements

You usually know how many material resources it takes to complete a task: In most cases, you can use a standard formula to calculate the number of pounds, tons, yards, or other quantity. But how do you know how much effort your work resources must invest to complete the tasks in the project?

TIP

Include a certain amount of extra material resources if you think you will need it for rework or scrap.

As with many aspects of information you put into a Project schedule, determining the level of effort rests to a great degree on your own experience with similar tasks and resources. Still, remember these caveats:

- **Skill counts.** A less skilled or less experienced resource is likely to take more time to finish a task. Experiment — increase the duration by 20 percent for a less skilled resource, for example, or reduce it by 20 percent for a more skilled resource.
- **History repeats itself.** Look at previously completed projects and tasks. If you've tracked people's time, you can likely see how much effort was required in order to complete various types of tasks on other projects and draw parallels to your project. This technique for estimating effort duration is similar to analogous estimating, covered in Chapter 5.
- **Ask and you shall receive.** Ask the resources themselves to estimate how long they think a task will take. After all, the people doing the work should know best how long it takes.

TIP

Allow for reserve time in a project to account for unforeseen circumstances, such as resources being less skilled than you had planned. To ensure that milestone dates are met, many people add a task called Contingency Reserve or Buffer to their schedules immediately before a milestone delivery or the end of a phase. See Chapter 12 for more information on contingency reserve.

# The Birth of a Resource

Just as someone fills out the birth certificate whenever a baby is born, a form is filled out whenever a resource in Project is born (that is, *created*). On the Resource Information form, you enter information such as the resource name, the availability, and the rate per hour or cost per use. You can also enter optional information, such as the group (company workgroup) to which the resource belongs or the resource's email address.

You can create a resource as a single person or entity, a generic resource (such as a skill set with no person assigned, such as Assistant or Engineer), or even a group of several resources that work together. Many times in the early stages of project planning, you know only the resource skill you need, to be replaced later by a named individual who has both the time and the qualifications to complete the task.

## Creating one resource at a time

On the simplest level, a resource is created as a single entity — a particular person or a meeting room or a piece of equipment. You create the resource by entering information in the Resource Information dialog box.

When you create a resource, you must type, at minimum, the resource name, though you can also add as much information as you want. Some people prefer to create all the resources first and add contact and cost information later. In Chapter 10, I describe in detail how to add the cost information for resources.

To create a resource using the Resource Information dialog box, follow these steps:

1. **Click the bottom of the Gantt Chart button on the Task tab in the View group of the Ribbon and click Resource Sheet.**

   At that point the view changes to the Resource Sheet view.

2. **Double-click a blank cell in the Resource Name column.**

   The Resource Information dialog box appears, as shown in Figure 7-2.

3. **Type a name in the Resource Name text box.**
4. **Click the down arrow in the Type box (on the right) to choose Work, Material, or Cost.**

   The settings that are available to you differ slightly, depending on which option you choose. For example, a material resource has the Email and Logon

Account options grayed out so they aren't available, and a work resource or cost resource has the Material Label option grayed out.

5. **For a material resource, enter a description of the units in the Material Label field.**

   For example, you might enter pounds for a Gravel resource or tons for a Steel resource.

6. **In the Initials field, type an abbreviation or initials for the resource.**

   If you don't enter anything, the first letter of the resource name is inserted when you save the resource.

7. **Continue to enter any information you want to include about the resource.**

   This information may include an email address, the group type (a department, division, or workgroup, for example), booking type (proposed or committed), or code (such as a cost-center code).

   If you enter information in the Group field, you can then use the Filter, Sort, and Group By features to look at sets of resources. (See Chapter 11 for more about filtering and working with groups.)

8. **Click the OK button to save the new resource.**

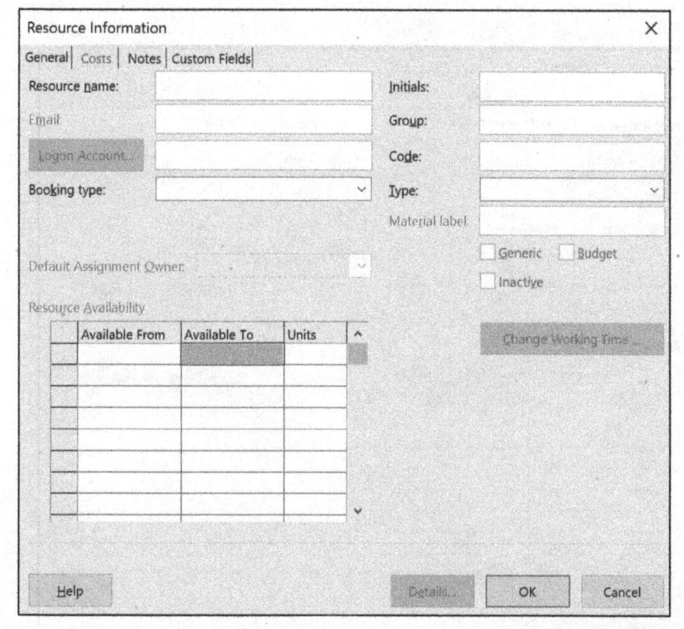


**FIGURE 7-2:** The Resource Information dialog box.

TIP

To enter information about several resources at a time, display Resource Sheet view and enter information in the columns. You can even paste resources copied from another source (such as an Excel spreadsheet) into this view.

## Identifying resources before you know their names

In the planning stages of a project, you often find that not all resources are assembled. Even well into the project, sometimes you don't know what resource you'll be using; you know only that you need a resource with a certain skill set to complete upcoming tasks. In this case, consider creating certain resources as generic resources.

To create a generic resource, give it a name that describes its skill, such as Engineer, Security Personnel, or even Meeting Space (rather than a specific resource named Conference Room B). Then, in the Resource Information dialog box, be sure to select the Generic check box, as shown in Figure 7-3. No formula exists for using the Generic setting to recalculate your schedule based on resource availability. However, many people find this setting useful when they use rolling wave planning or when they aren't responsible for specific resource assignments (for example, assigning a temporary worker to a task when a temporary-hiring agency will choose the specific worker). For a refresher on rolling wave planning, see Chapter 3.

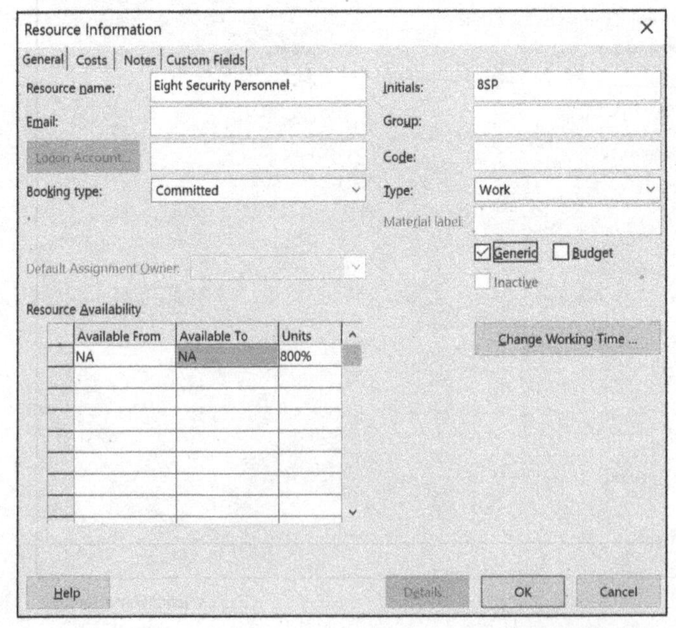


**FIGURE 7-3:** Generic consolidated resource.

© John Wiley & Sons, Inc.

You can replace a generic resource name with the name of the resource that's added to your project. For example, if you enter **Security Lead** as a resource, you can replace it with **Ty Mack**, the name of the security professional who will do the work. To make the switch, go to Resource Sheet view and, in the Resource Name column, click the cell for the resource, type the new name, and press Enter. Don't worry if you've already assigned the generic resource to tasks in the project. Project updates all assignments with the new name.

## Many hands make light work

Rather than assign people individually to a task, you can assign a group of people who typically work together. Being able to make a single assignment of a *consolidated resource* rather than several separate resources and assignments can be a time-saver in larger projects.

Suppose that you need eight security personnel for the Youth Center event, and you create a Security Personnel resource. This functions as a consolidated resource that represents eight individual resources. In the Resource sheet, you enter **800%** in the Max. column to indicate eight full-time resources.

No special setting is used to designate a multiple resource. However, you may want to indicate the number of resources by adding it to the resource name. For example, name your security personnel resource Eight Security Personnel. The defining factor in this type of resource is the maximum number of assignment units: 800% signifies eight resources in the group. Figure 7-3 shows Security Personnel with the Generic check box selected and the units at 800%.

TIP

Assigning a resource at less than 100 percent units creates a part-time assignment. Of course, another option is to keep the resource in the pool at 100 percent and assign them only part time to a task. It's your choice.

# Managing Resource Availability

Lots of Project features handle resources — in particular by helping you spot resource overallocation. *Overallocation* is a calculation, based on the resource's calendar and availability, indicating that too much work has been assigned based on availability.

As an example, consider Damian Van Pelt, an electrician who works a standard eight-hour day. Damian is assigned at 100 percent of his availability to the Install Card Readers and the Install Infrared Photo devices tasks, which occur at the same time. Damian is now working at 200 percent of his availability, or 16 hours per day. Poor Damian is overbooked. Be careful — he might even quit the project.

A resource is assigned to a task at 100 percent availability (or units) by default, but you can modify that setting if you know that a resource will be assigned to several tasks and can only spend a limited amount of time working on each one.

## Estimating and setting availability

Availability is easier to estimate for some resources than for others. Managers are unlikely to devote an entire day to any single task, because they have to deal with all the employees who report to them and also sign authorizations, attend meetings concerning various projects, and develop budgets, for example. A production worker's availability may be simpler to pin down to a single task: If one manufacturing job spends three days on the line and one person is working on the line the entire time, it's more accurate to say that they are working on that task full time.

One mistake made by Project beginners is to overthink availability. Of course, no one spends eight hours every day on a single task in a project. Employees spend part of their workdays reading email, chatting with coworkers, and answering phone calls about issues unrelated to their projects. A resource may spend seven hours on a task one day and only three hours the next. Don't get hung up on a day-by-day resource schedule when estimating availability. If, over the life of a task, the employee focuses on it, 100 percent availability is an appropriate setting. If that person works only five days on a ten-day task, however, that's 50 percent availability, whether they work four hours per day for ten days or five full days at any point.

REMEMBER

The Units (availability) setting helps you spot the overbooking of a resource who may work on multiple tasks at the same time in a project schedule.

To set default resource units to specify availability, follow these steps:

1. **Display the Resource Sheet view.**
2. **Click the Max column for the resource.**
3. **Type a number representing the percentage of time that the resource is available to work on the project.**

   For example, type 33 for a resource who's available one-third of the time or 400 for a resource who can provide four full-time workers. (Entries greater than 100 percent represent resources that provide a group or team to handle assignments.) The most common entry (the default) is 100 for a single resource working full time on assignments in the project.

4. **Press Enter or Tab.**

   The entry is completed.

Note that this only works for resources whose type is set to Work.

## When a resource comes and goes

In addition to being available to a project only a certain percentage of the time (say, 50 percent of the project's entire duration), a resource may be available for only a certain period of time (perhaps April–June) during the life of the project — commonly, when you use a resource borrowed from another department or when you're working with a freelancer who's squeezing in your project with others.

Another possible resource is one who's available half time for the first few days of the project and then full time for the remainder. In this case, you enter a date range in the Available From and Available To columns of the Resource Availability area in the Resource Information dialog box, as shown in Figure 7-4, to specify varying availability.

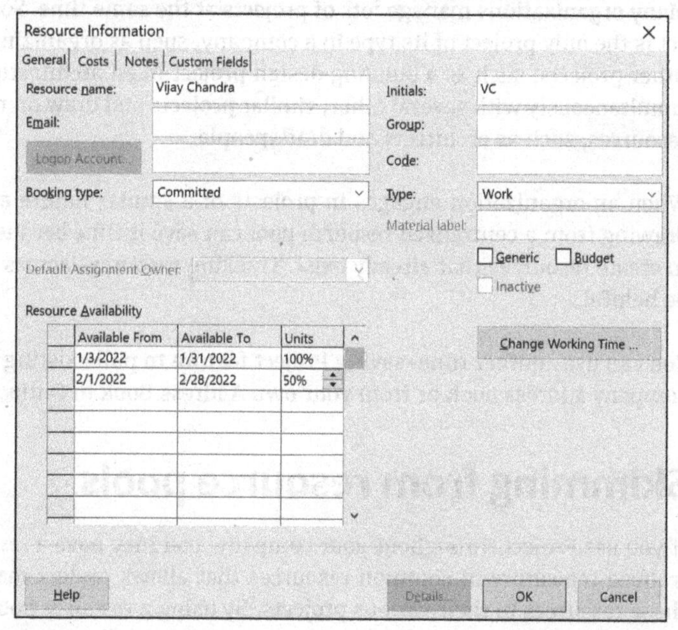


**FIGURE 7-4:** Time periods and units of availability.

To specify a limited period during which a resource is available to work on your project, follow these steps:

1. **Display the Resource Sheet view.**
2. **Double-click a resource.**

   The General tab of the Resource Information dialog box appears (refer to Figure 7-4).

3. **Use the Available From and Available To columns (in the Resource Availability area) to specify the period of availability.**
4. **In the Units column in the same row, either click the arrows to raise or lower the availability in 50 percent increments, or type a number.**

   Press Enter to finish the entry.

5. **Repeat Steps 3 and 4 to specify additional periods of availability on subsequent rows in the Resource Availability area.**
6. **Click the OK button to save the changes.**

# Sharing Resources

Many organizations manage lots of projects at the same time. Sometimes, a project is the only project of its type in a company, such as organizing an office move. Other projects, such as a building design project in an architectural firm, happen simultaneously with several other, similar projects and draw on many of the same resources, such as architects and draftspeople.

When an organization engages in projects of a similar nature at the same time, drawing from a centralized resource pool can save it time because it doesn't have to create resources that already exist. Tracking resources across projects can also be helpful.

You can use another time-saving Project feature to pull existing resources from a company address book or from your own Address Book in Outlook.

## Skimming from resource pools

If you use Project throughout your company, you may have a *resource pool*, a centralized repository of common resources that allows project managers to assign those resources to their various projects. By using a resource pool, you can form a more realistic idea of how busy resources are across all projects at any time.

## CREATING SHARER FILES

Both individual resources and generic resources can be created in a blank project as a resource pool and saved to an accessible location on your company network. Then any project manager can assign those resources to their own projects. These projects are then referred to as *sharer files* because they share resources from the resource pool. For example, if you have a pool of maintenance people whom everyone in your manufacturing company assigns to projects, create a Project file named `Resource Pools` and enter all the resources to share in Resource Sheet view. Or create the resource pool file named CEO and let everyone who is managing projects that require the CEO's involvement and assign them from that central location. Then use the resource-sharing tools in Project to assign these resources to tasks within your plan.

When anyone uses resource assignments from a sharer file, that information is also saved in the resource pool file. Then anyone can use that file to look at resource allocations across all projects in the organization.

TIP

Don't confuse resource pools with *enterprise resources*, which require that you set up Project Professional, Project Server, and Microsoft Office Project Web App. A resource pool is simply a Project file that contains only a list of resources in Resource Sheet view and is saved on a shared network drive for your workgroup or on your company server. Anyone with access to the resource pool file can assign the resources it contains to projects. A resource pool saves everybody the trouble of having to repeatedly create these resources their individual projects.

To access a resource that's available to your entire organization, follow this procedure:

1. **Open the resource pool file and then open the file that will be the sharer file.**

   See the earlier sidebar "Creating sharer files" for information about the resource pool and sharer file.

2. **Working in the sharer file, choose Resource⇨Resource Pool⇨Share Resources.**

   The Share Resources dialog box appears.

3. **Specify the resources for the project.**

   If you want to specify that a project will use only its own resources (the default setting), select the Use Own Resources option. If you want to share resources, select the Use Resources option and then choose a project from the From list.

4. **Specify what Project should do when a conflicting resource setting, such as the resource base calendar, exists.**

   If your project's setting will take precedence, select the Sharer Takes Precedence option. If you want the pool setting to rule, select Pool Takes Precedence.

5. **Click the OK button to complete the process.**

   All resources in the specified resource pool are added to your own project's resource list, ready to be assigned to tasks.

After you add a shared resource to your project, you can update shared resource information, in case the person who maintains those shared resources has made a change, such as increasing the resource's rate per hour. To do this, choose Resource⇨Resource Sharing⇨Refresh Resource Pool.

WARNING

If you combine separate projects into one master project at any point, Project allows you to have duplicated resources. If you link the combined projects and then delete a duplicate resource in the master project, it's deleted in the subproject as well. A better practice is to have all the separate projects share resources from a resource pool so that no duplicate resources exist when you create the master project file.

## DROWNING IN THE RESOURCE POOL

Drawing resources from resource pools saves you time because you don't have to repeatedly re-create those resources. However, you may wonder whether you should track your resource's time in the resource pool file to see whether the resource is overbooked. Most projects in the real world use resources that aren't solely dedicated to a single project. New users of Project are often confused because almost every person working on their projects spends time on other work, from general communication with coworkers and clients to efforts invested in other projects. Should these beginners build resource pools to account for time shared among several projects at one time?

Generally speaking, trying to track every minute of every resource's day to see whether resources are working 100 percent or 50 percent on your tasks or are being shared among multiple projects amounts to chaos. Ask yourself this question: When this resource works on a task in your project, will they put their entire focus on that task at that time? If so, you may not need to fool around with tracking shared resources across many projects. Especially on shorter tasks, resisting the urge to micromanage the efforts of your resources outside your own project usually works well. If, on the other hand, you have resources working only half the time or splitting time between two projects routinely, consider using shared resource tools to keep track of those resources across projects.

## Importing resources from Outlook

Project allows you to save yourself the time spent entering resource information by allowing you to pull resources from Outlook.

To pull resources from Outlook, you must have it specified as your default email program: Open Outlook, and when you're asked whether you want it to be your default program, choose yes.

When you insert one or more Outlook resources into your project, they're added to your project list, using the resource name and email address as they exist in the Outlook Address Book. The default first-letter initial and work type are also pre-assigned. You can then add to the resource any details you like.

To insert resources from your Outlook Address Book, display Resource Sheet view and follow these steps:

1. **Go to the Resource Tab on the Ribbon. In the Insert group, click Add Resources, and then click Address Book.**

   The Select Resources dialog box appears.

2. **Click the name of a resource in the Name list.**
3. **Click Add to place the selected name in the Resources list.**
4. **Repeat Steps 2 and 3 to add to your project all the resource names you want to import.**
5. **When you are finished, click the OK button.**

The names now appear in the project resource list, ready for you to assign them to tasks.

IN THIS CHAPTER

» Exploring the base, project, resource, and task calendars

» Understanding how calendars work together

» Setting calendar options and working times

» Choosing the project calendar

» Working with task calendars

» Using resource calendars

» Creating your own calendar templates

» Copying calendars to another project

# Chapter 8 Working with Calendars

Most people live their lives based on responding to their clocks and calendars. Think about it: You wake up, and your first thoughts are about what day it is, what time it is, and whether it's a workday.

You have a typical workday, whether you're a 9-to-5 person or your particular job calls for you to work from midnight to 8 a.m. You probably also vary from that routine now and then, when you complete a 12-hour marathon session in a crunch or slip away after half a day to go fishing on a nice summer day.

In a way, Project calendars resemble your work life: They set standards for a typical working time and then allow for variation. Unlike you, however, Project has several types of calendars to account for.

## Mastering Base, Project, Resource, and Task Calendars

One piece of information to consider when assigning resources is *when* people can work. You adjust resource working time in calendars.

### Setting the base calendar for a project

As I describe in Chapter 2, when you first open a new project, you enter information about the project either in the Project Information dialog box or by clicking the File tab and working in the Project Information section of the Info screen in Backstage view. One option under Project Information is the Project Calendar.

You can choose among three options for the base calendar:

- The **Standard** (default) calendar shows a typical 8-to-5 workday and a 5-day workweek, with no exceptions such as holidays. You can think of it as the company calendar, which reflects its operating hours (when the "doors are open").
- **Night Shift** shows an 8-hour workday, scheduled between 11 p.m. and 8 a.m., with an hour off for a meal, from Monday through Friday — no exceptions. You can select this calendar for resources who work the same company days but different hours.
- The **24 Hours** base calendar shows continuous time, 24 hours per day and 7 days per week.

### Understanding the four calendar types

Project has four calendar types: base, project, resource, and task. Mastering the calendar types can be tricky, but understanding how calendars work is essential to mastering the software. Tasks are scheduled and resources are assigned based on the calendar settings you make.

Here's the lowdown on the role of each of the four calendar types in Project (the next section details how they interact):

- **Base:** The template on which all other calendars are built. Three base calendars are available: Standard, Night Shift, and 24 Hours. (You can read more about them later in this chapter, in the section "Working with Task Calendars and Resource Calendars.")

- **Project:** The default calendar for scheduling. You choose which base calendar template your particular project should use.
- **Resource:** Combines the base calendar settings with any exceptions (nonworking times, such as vacation time) that you set for a particular resource.
- **Task:** Provides a spot where you can set exceptions for a particular task.

You can manage calendars in a project in one of three ways: Use a project calendar that sets the working dates and times for all resources and tasks, modify calendars for specific resources, or assign and modify calendars for specific tasks.

REMEMBER

Don't become confused by my use of the term *project* as it refers to the software, the project calendar type, and your project. I capitalize *Project* to indicate the software itself, *project calendar* to specify the calendar type, and plain *project* (lowercase) to refer to your project.

## How calendars work

Not everyone in a company works the same schedule, and not every task can be performed in the same 8-hour workday. Nevertheless, when you create tasks and assign resources to work on them, Project has to base that work on a timing standard. For example, if you say that a task should be completed in one workday, Project knows that a *workday* means 8 hours (or 12 hours or whatever) because that's how you set up a standard workday in the project calendar. Likewise, suppose that you assign a resource to work two weeks on a task in a company that uses a standard five-day workweek. If that resource's own calendar is set for a standard four-day workweek, the two weeks of work by that resource defer to the timing of the resource calendar for a total of only eight workdays. In other words, the resource calendar overrides the standard calendar.

WARNING

The nature of a task can affect resource time. A two-week, effort-driven task isn't complete until its resources have invested two weeks (according to the project calendar or task calendar) of effort. You can find out more about effort-driven tasks in Chapter 5.

## How one calendar relates to another

All calendars in a project are controlled, by default, by the project calendar setting. Here's the tricky part, though: When you change a task calendar or resource calendar, Project creates an exception to the project calendar. Thus, to have an accurate schedule you have to understand which setting takes precedence.

Here's how the precedence concept works:

- With no other settings made, the base calendar template you select for the project calendar when you first create the project controls the working times and days of all tasks and resources.
- If you make changes in the working hours for a resource, those settings take precedence over the project calendar for that resource when you assign it to a task. Likewise, if you assign a different base calendar for a task, that calendar takes precedence over the project calendar for that task.
- If you apply one calendar to a resource and a different calendar to a task that the resource is assigned to, Project uses only common hours to schedule the resource. For example, if the task calendar allows work from 8 a.m. to 5 p.m. and the resource calendar allows work from 6 a.m. to 2 p.m., the resource works from 8 a.m. to 2 p.m., which is the only period the calendars have in common.
- You can set a task to ignore resource calendar settings by opening the Task Information dialog box (double-click the task name in Gantt Chart view) and selecting the Scheduling Ignores Resource Calendars check box on the Advanced tab. (This setting isn't available if the task calendar is set to None.) You might make this setting if you know that all resources are required to be involved in a task (such as a quarterly company meeting), regardless of their usual work hours.

# Scheduling with Calendar Options and Working Times

In this section, I throw two more timing elements at you: calendar options and working times.

*Calendar options* are used to change the standard settings for a working day, week, and year. You can set a project calendar to Standard, for example. By default, that's 8 a.m. to 5 p.m., five days a week.

*Working time* is used to adjust the time available for work on a particular date or days. Suppose that you make a change to the calendar options so that you have 8-hour days and 32-hour workweeks. Also check your working time to ensure that you specify 3 nonworking days of the 7-day week, to jibe with the 32-hour week. If you want to set a certain date as *nonworking* for your project, such as the company offsite meeting day, you can do that using the working-time settings.

TIP

To ensure that the project is scheduled accurately from the get-go, make all changes to calendar options and working times *before* you add tasks to the project.

WARNING

You have to change the calendar options and calendar settings for *every* new project file you create. However, you can set up a calendar that can be used for all your projects and shared in a single global location, as explained later in this chapter, in the section "Sharing Copies of Calendars."

## Setting calendar options

When you make changes to a resource calendar or task calendar, you simply adjust the times that a resource is available to work or the time during which a task occurs; you don't change the length of a typical workday for the project. A day is still 8 hours long if that's the project calendar setting, even if you say that a task that takes place on that day uses the 24-hour base calendar template.

To change the length of a typical workday to 10 hours rather than 8, for example, you must do so on the Calendar tab of the Options dialog box.

Follow these steps to modify the calendar options:

1. **Select the File tab on the Ribbon and click the Options button.**

   The Project Options dialog box appears.

2. **Click the Schedule category on the left side of the Project Options dialog box, as shown in Figure 8-1.**
3. **From the Week Starts On drop-down list, choose a day.**
4. **To modify the start of the fiscal year, select a month from the Fiscal Year Starts In drop-down list.**
5. **To change the working hours in a typical day, type new times in the Default Start Time and Default End Time fields.**

   REMEMBER

   If you change the default start or end time setting, change the corresponding working time also, as described in the following section.

6. **Modify the Hours Per Day, Hours Per Week, and Days Per Month fields as needed.**
7. **Click the OK button to save these settings.**

REMEMBER

Worrying about every calculated hour can be tedious. Most professional schedulers avoid changing the default calendar settings. Unless you truly want to track the scheduling of tasks to the hour, simply manage tasks to the day or week and keep track of scheduled deliverables in this manner.

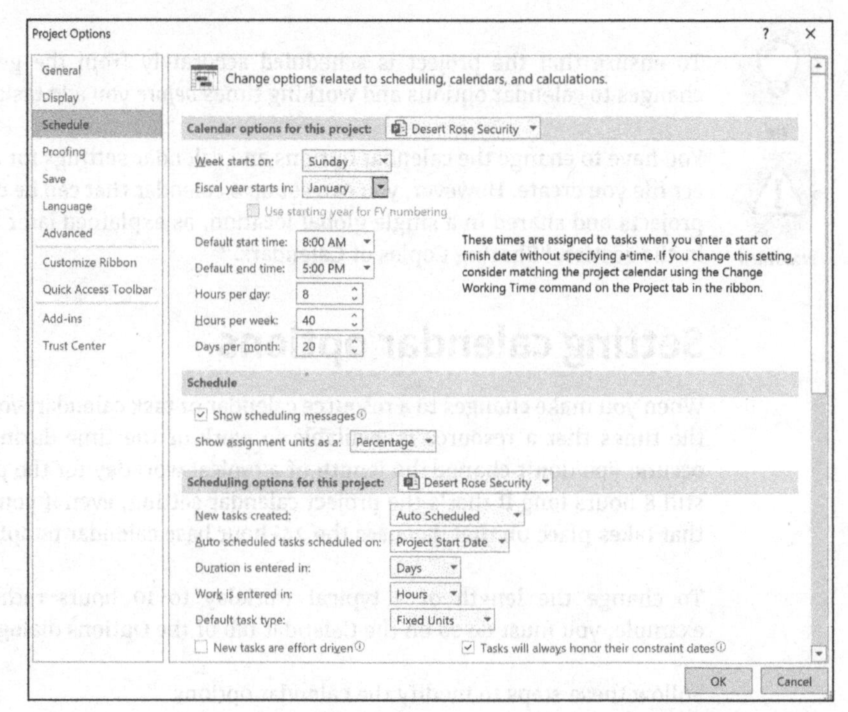


**FIGURE 8-1:** Define typical work times.

*© John Wiley & Sons, Inc.*

## Setting exceptions to working times

To change the available working hours for a particular day (such as December 24), you use the working-time settings. For example, if the day before Thanksgiving is half a workday, you can modify the working-time settings for that day; then any resources assigned to a task on this date spend only half a day working. You also use these settings to specify global working and nonworking days to match the calendar options settings.

Here's how to change working times:

1. **On the Project tab, go to the Properties group and click the Change Working Time button.**

   The Change Working Time dialog box opens, as shown in Figure 8-2.

2. **On the calendar, click the day you want to change.**
3. **Click the Exceptions tab to display it; then click the Name cell in a blank row, type a name for the exception, and press the Tab key.**

   The Start and Finish dates will autopopulate with the date you selected in Step 2.

4. **Click the exception name you entered in Step 3 and then click the Details button.**

The Details dialog box for this exception opens, as shown in Figure 8-3. This example shows Thanksgiving Eve with the workday ending at 2 p.m.

5. **Select either the Nonworking radio button or Working Times radio button.**
6. **Enter a time range in the From and To fields.**

TIP

To set nonconsecutive hours (for example, to build in a lunch break), you have to insert two or more sets of numbers (such as 8 to 12 and 1 to 5) in the From and To fields.

7. **Select a recurrence pattern and then set an interval in the Recur Every *x Time Period* field.**

   For example, if you select Weekly and click the arrows to set the interval field to 3, this pattern recurs every three weeks.

8. **Set the range of recurrence.**

   Either enter Start and End By dates or select the End After radio button and set the number of occurrences there. This example ends after one occurrence.

9. **Click the OK button to close the Details dialog box and click OK again to close the Change Working Time dialog box. Save your changes.**

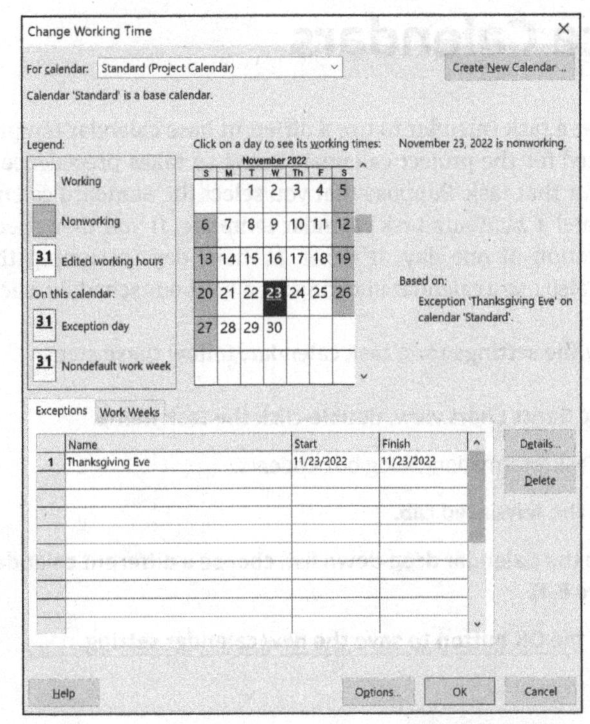


**FIGURE 8-2:** The standard calendar with default working times.

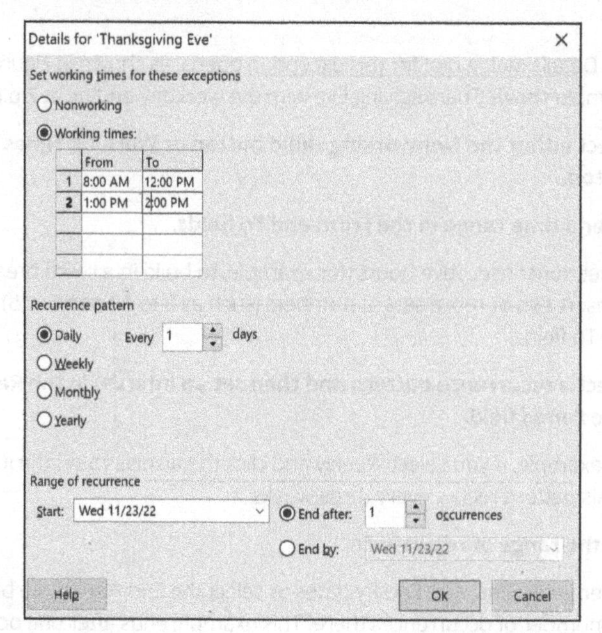


**FIGURE 8-3:** Modifying the default calendar.



# Working with Task Calendars and Resource Calendars

You can set a task calendar to use a different base calendar template from the one you selected for the project calendar. Doing so takes precedence over the project calendar for that task. Suppose that you select the Standard calendar template for a project and a 24 Hours task calendar template. If you then specify that the task has a duration of one day, it's one 24-hour day (assuming that the assigned resources also use a calendar allowing for a 24-hour schedule, such as machinery).

To modify the settings for a task calendar, follow these steps:

1. **In the Gantt Chart view, double-click the task name.**

   The Task Information dialog box appears.

2. **Click the Advanced tab.**
3. **From the Calendar drop-down list, choose a different calendar (see Figure 8-4).**
4. **Click the OK button to save the new calendar setting.**

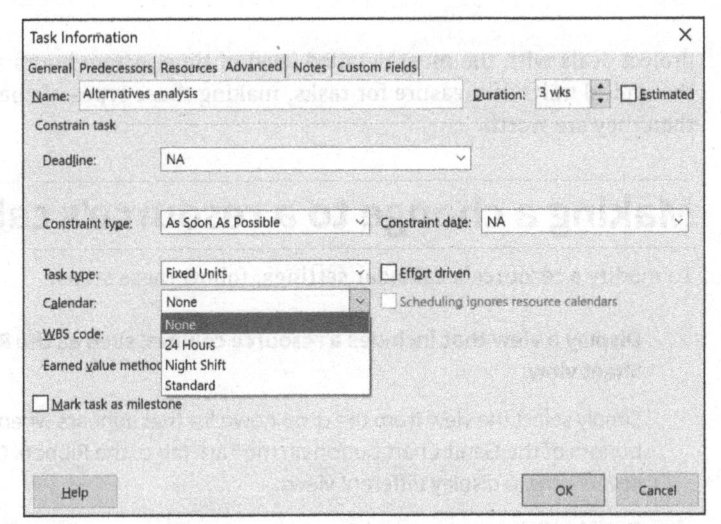


**FIGURE 8-4:** Changing the task calendar.

© John Wiley & Sons, Inc.

REMEMBER

If a resource assigned to a task has a modified calendar, the resource works only during the specific hours that the task calendar and resource calendar have in common.

## Setting resource calendars

Only work resources have their own calendars. That's because material resources are charged not by time worked but by units used, and a cost resource is assigned a set cost that doesn't relate to any time worked on a task.

Even the most resourceful resources have only so many hours in a day to work. When you have to deal with variations in resource schedules, consider modifying the resource calendars.

You can change the base calendar template for each work resource and set specific dates as working or nonworking. These exceptions take precedence over the project and task calendars, and control when a specific resource can work.

REMEMBER

Unless a resource has a truly unique working schedule, don't change its base calendar template. For example, if a resource usually works a day shift but works a night shift for only a few days during the life of the project, don't change that resource's base calendar template to Night Shift. If one person works from 10 a.m. to 7 p.m. because the company allows them to, you probably don't have to vary their schedule from the typical 8-to-5 work schedule that's set in the project calendar, because they work eight hours a day, like everyone else does. Unless your project deals with the most detailed level of time, where hours and not days are the typical units of measure for tasks, making these types of changes more work than they are worth.

## Making a change to a resource's calendar

To modify a resource's calendar settings, follow these steps:

1. **Display a view that includes a resource column, such as the Resource Sheet view.**

   Simply select the view from the drop-down list that appears when you click the bottom of the Gantt Chart button on the Task tab of the Ribbon. Chapter 6 covers how to display different views.

2. **Double-click a resource name.**

   The Resource Information dialog box appears.

3. **Click the Change Working Time button on the General tab to display the Change Working Time dialog box.**

   The Exceptions and Work Weeks tabs have settings that are identical to the ones in the Change Working Time dialog box for tasks, but changes made here affect this resource rather than the task.

4. **Click on a day on the calendar to see its working times.**
5. **Click the Work Weeks tab to display it; then click a blank row, enter a name for the exception, and press Enter.**

   The Start and Finish dates populate with the date you chose. If you want to edit the default workweek for all weeks of the year, leave [Default] selected instead. If you want to mark a single holiday, use the Exceptions tab instead.

6. **Click the exception you just created, and click the Details button.**

   The Details dialog box for this exception appears.

   Figure 8-5 shows the Details dialog box.

7. **Select the day or days to modify in the Select Day(s) list on the left side.**

   You can click the first day and then Shift+click to select a range of adjacent days, or Ctrl+click to select nonadjacent days.

8. **To set working times for these exceptions, select either the Nonworking radio button or Working radio button. If you chose nonworking time, click OK.**

9. **If you chose to set specific working times in Step 8, enter a time range in the From and To fields. Click OK.**

   To delete a row in the From and To area, click the row number and press Delete.

REMEMBER

   To set nonconsecutive hours (for example, to build in a lunch break), put two or more sets of numbers (such as 8 to 12 and 1 to 5) in the start and finish columns.

10. **Set the range of recurrence.**

    Change the Finish date to meet your needs, if needed.

    Figure 8-6 shows how a resource's vacation is shown in the Change Working Time dialog box.

11. **Click the OK button to close the Change Working Time dialog box and save these changes.**

12. **Close the Resource Information dialog box.**

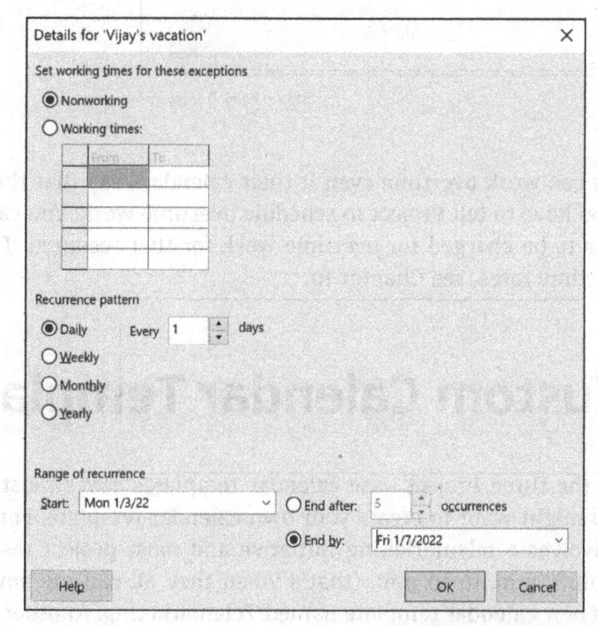


**FIGURE 8-5:** The Details dialog box.

© John Wiley & Sons, Inc.

WARNING

Micromanaging nonworking time for your resources can leave you no time to do anything else, so avoid this urge. For example, if someone takes off half a day for a doctor's appointment, don't block off a day. However, if a resource takes a two-week vacation or a three-month sabbatical, you should probably modify that resource's calendar.

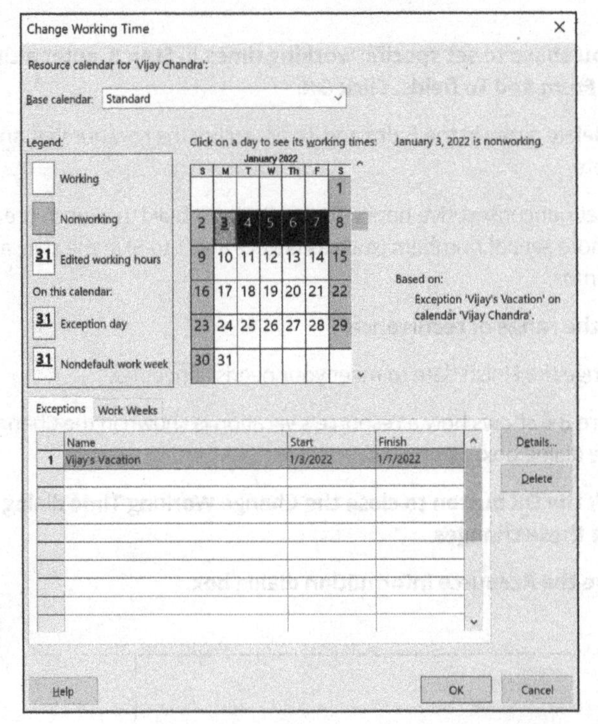


**FIGURE 8-6:** Changing working time on a resource calendar.

REMEMBER

Resources can work overtime even if their calendars say that they're 8-to-5ers, though you have to tell Project to schedule overtime work. You can also set a different rate to be charged for overtime work for that resource. To find out more about overtime rates, see Chapter 10.

# Creating a Custom Calendar Template

Although the three Project base calendar templates cover most working situations, you might want to create your own calendar template. For example, if the project involves a telemarketing initiative and most project resources work six hours, from 4 p.m. to 10 p.m. (that's when they all call me anyway!), consider creating a new calendar template named Telemarketing. Another reason to create a custom template is that the three Project calendars have no holidays specified. You may need to ensure, at minimum, that company holidays are indicated in the calendar you use to ensure that Project accurately schedules around those dates.

If you want to save some time when creating a template, start with an existing base calendar template that most closely fits your needs. Then modify it as you like by making changes to the working times and calendar options (see the section "Scheduling with Calendar Options and Working Times," earlier in this chapter) to ensure that they're in agreement. After you create a new calendar template, it's available for you to apply in all three calendar types: project, task, and resource.

Because the project calendar is the basis of the entire project, it should represent the most common working schedule in the project. If only certain resources in the project work odd hours, change the resource calendars and not the project calendar.

Follow these steps to create a new calendar template:

1. **Click the Project tab, go to the Properties group, then click Change Working Time.**

   The Change Working Time dialog box appears.

2. **Click the Create New Calendar button.**

   The Create New Base Calendar dialog box appears, as shown in Figure 8-7.

3. **In the Name box, type a unique name for the new calendar.**
4. **Select either the Create New Base Calendar radio button or the Make a Copy of *x* Calendar radio button; then select an existing base calendar from the list on which to base the calendar template.**

   If you choose Create New Base Calendar, Project creates a copy of the Standard calendar with a new name. If you choose Make a Copy Of and select 24 Hours or Night Shift, the new calendar is based on that choice. Whichever you choose, it's your starting point, and you can make changes to make the calendar unique after making this choice.

5. **Click the OK button to return to the Change Working Time dialog box.**

   Now make changes to the working time for the new calendar template.

6. **Click the OK button to save the new calendar settings.**

If you haven't already done so, change the calendar options to match the working times of the custom calendar you created. You can do this from the Change Working Time dialog box, using its Options button.

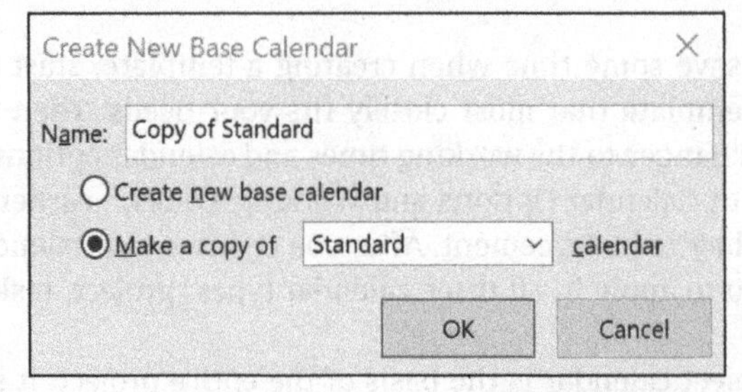


**FIGURE 8-7:** Creating a new calendar.

© John Wiley & Sons, Inc.

# Sharing Copies of Calendars

You can make a calendar available to multiple projects by using the *Organizer*; it serves as the control center for custom items such as calendars you create in your project files. Using the Organizer, you can copy a custom calendar from one project file into another. Or you can copy a custom calendar into a template that provides default settings to every new, blank project file you create.

To copy a calendar from one project to another, follow this procedure:

1. **Open the project to which you want to copy a calendar and the project file that holds the custom calendar to copy.**
2. **Choose File ⇨ Info.**

   Backstage view opens.
3. **Click the Organizer button.**

   The Organizer dialog box appears.
4. **Click the Calendars tab.**
5. **From the Calendars Available In drop-down list (in the lower-left corner), select the Project file that contains the calendar you want to copy.**
6. **In the Calendars Available In drop-down list (in the lower-right corner), choose whether you want to make the calendar available in another open project (by selecting that project) or the Global template (by selecting `Global.MPT`).**
7. **In the list on the left, click the calendar you want to copy and then click the Copy button, as shown in Figure 8-8.**

   The calendar is copied to the current project.

8. **If you want to give the calendar you copied a different name, make sure that it's selected in the list, click the Rename button, type a new name in the Rename dialog box that appears, and then click the OK button.**

9. **Close the Organizer by clicking the X (Close) button in the upper-right corner.**

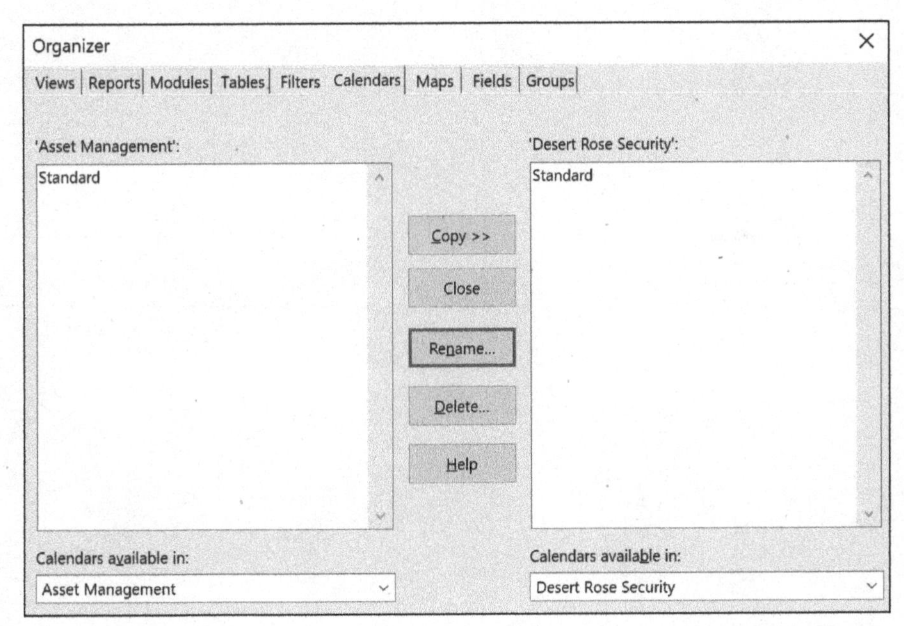


**FIGURE 8-8:** Copying a calendar to other projects.

© John Wiley & Sons, Inc.

TIP

Follow these pointers to copy calendars from project to project:

- **Give the calendar a descriptive name.** Providing an appropriate name helps you remember the calendar's general parameters.
- **If your company has standard calendars, have a single resource create and disseminate them.** If ten versions of a management calendar float around and you grab the wrong one, it can cause problems.
- **Add the project manager's initials to every calendar template name you create.** That way, you know which ones you created.

IN THIS CHAPTER

» Identifying the right resource

» Making resource assignments

» Checking resource availability

» Using Team Planner view

# Chapter 9
# Assigning Resources

After you enter tasks and resources, you start assigning tasks to resources. That is when the fun begins! The duration of some tasks may change, and you may start to see evidence of people who are overbooked on several tasks that happen around the same time. Understanding how these results occur is the key to making intelligent assignments and balancing the project scope with your milestone obligations and resource availability.

In fact, assigning resources, balancing scope, schedule and cost, and replanning is an ongoing process throughout a project. As usual, Project provides the tools to help you manage the resource assignment process so you can arrive at an optimal solution.

## Finding the Right Resource

Sometimes, no one in the world can perform a certain task like Albert does, and you'll persuade Albert to do that job if it kills you. At other times, almost anyone can handle the work.

If any Tom, Dick, or Mary with a certain skill level (or a certain rate per hour) can do the job, you can use Project features to find the right resource and ensure that they have enough time to take on "just one more" task.

## Needed: One good resource willing to work

You've probably used the Find feature in other software to find a word or phrase or number. That's child's play compared with Project's Find feature, which can find you a backhoe, a corporate jet, or even a person! You can use Project's Find feature to look for resources with certain rates or in a certain workgroup. You can search for resources by their initials, their maximum assignment units, or their standard or overtime rate.

Suppose that you need to find a resource whose standard rate is less than $60 per hour. Or you want to find someone who can spend extra time working on a task, so you search for any resource whose maximum units are greater than 100 percent. (In other words, the resource can work a longer-than-usual day before they are considered *overallocated.*) Perhaps you need to find a material resource that is a chemical measured in gallons, but you can't remember its exact name. In this case, you can search for resources whose material label includes the word *gallons.*

REMEMBER

To find resource information, you must be working in a resource-oriented view. To find task information, you must be working in the task-oriented view.

First, display any resource view and then follow these steps to find resources in Project:

1. **On the Task tab, click Find in the Editing group.**

   The Find dialog box appears, as shown in Figure 9-1.

2. **In the Find What box, type the text you want to find.**

   For example, type 75 if you want to search for a resource with a standard rate of $75 per hour or less, or type Rebar if you want to find a resource whose material label contains that word.

3. **From the Look in Field list, choose the name of the field in which you want to search.**

   The default field is Name. When you click the drop-down arrow, you see all kinds of different fields you can search. For example, to search for resources that have a Standard Rate of $60 or less, choose the Standard Rate field.

4. **In the Test box, select a criterion.**

   The default value is Contains, but when you click the drop-down arrow, you see lots of options. For example, if you're looking for an hourly rate of $60 or less, try Is Less Than or Equal To.

5. **If you prefer to search backward from the current location (that is, the selected cell in the task list) instead of forward, choose Up from the Search list.**

6. **If you want to match the case of the text, select the Match Case check box.**
7. **To begin the search, click the Find Next button.**
8. **Continue to click Find Next until you find the resource you're looking for.**

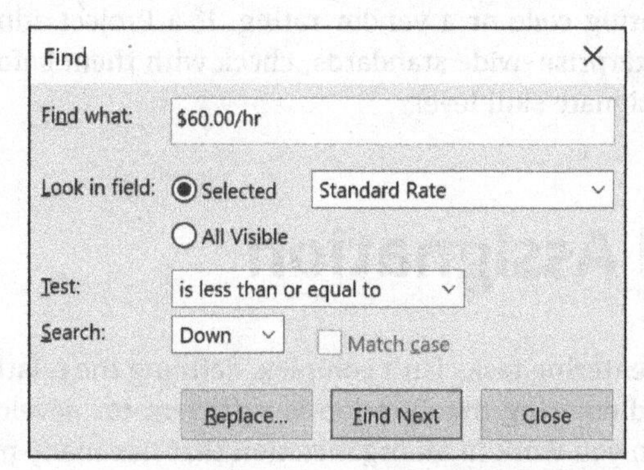


**FIGURE 9-1:** Using the Find feature.



## Custom fields: It's a skill

When assigning resources, you often need to consider a person's skills. If a person with less skill or experience could work on a particular task (and save you money because they charge a lower rate per hour), wouldn't it be nice to be able to find these resources easily?

Well, Project doesn't include a Skill field, but it does allow you to add fields of your own. You can use these fields for anything, but one useful way is to code resources by skill level. You can use a rating system such as A, B, and C or use terms such as *Exp* for an experienced worker and *Beg* for a beginning-level worker.

Here's how to add a custom field:

1. **Display the Resource sheet (or whatever sheet in which you want to view the custom field).**
2. **Scroll to the right on the sheet until you get to the end of the columns and click the Add New Column heading.**
3. **Type a name for the field and press Enter.**

You can enter whatever you like in this column for each resource in the project. Then you can search for specific entries in that field using the Find feature or turn on a filter to display only resources with a certain skill level in that field. (Read more about filters in Chapter 11.)

TIP

The custom field that's created using this method uses one of the placeholder fields Text 1 through Text 30 that's available for you to customize. You can customize other types of placeholder fields, such as cost fields or flag fields. To access more custom fields, right-click a field column heading and click Custom Fields.

WARNING

Some organizations designate custom fields for certain company information, such as an accounting code or a vendor rating. If a Project administrator is in charge of these enterprise-wide standards, check with them before you choose a custom field to designate skill level.

# Making a Useful Assignation

In a project where entering tasks isn't complex, defining the relationship, assigning a resource, and entering the duration is sufficient for developing a working schedule. But when you work in an organization that has many projects competing for limited resources and those projects are complex in nature, you're better served using software to help you manage its complexities.

Alas, the more complex the environment, the more variables that affect the schedule. For example, you can't simply enter a series of tasks; you need to determine the type of task (fixed units, fixed work, or fixed duration) and determine whether the task is effort driven. The task type affects the duration of the task after you assign resources. I describe task types in Chapter 5.

When you assign resources, pay attention to the resource type (work, material, or cost) and the resource availability. In addition, consider whether the project calendar, task calendar, and resource calendars affect the tasks assigned to resources. You can use a couple of methods to assign resources to tasks and also specify the resource assignment units. These assignment units differ slightly between work resources and material and cost resources, as explained in the next several sections.

## Determining material and cost-resource units

Work resources, which are typically people, are assigned to a task using a percentage: for example, 50 percent or 100 percent. When you assign a resource at a percentage, the assignment is based on the resource calendar. A resource with a Standard calendar spends eight hours a day working if you assign it at 100 percent assignment units.

A *material resource* is assigned in units, such as gallons, consulting sessions, yards, or tons. When you assign a material resource to a task, you designate how many units of that resource are devoted to that task. Project doesn't act like job-costing software, deleting units from an inventory; it isn't affected by how many units are used against the total available. (You can't use it in this manner.)

A *cost resource* is one that incurs a variable cost every time you assign it. For example, if you create the Permit Fee cost resource, every time you assign a permit fee to a task, you can designate the actual cost for that assignment, whether it's $25 or $125.

# Making assignments

You have three main ways to assign resources in Project (although you can use other methods while working in various views):

- Select resources from the Resources tab of the Task Information dialog box.
- Enter resource information in the Resource Names column in the Entry table (displayed in Gantt Chart view).
- Use the Assign Resources dialog box.

Regardless of which method you use, you work in a task-based view, such as Gantt Chart view, to make the assignments.

Which method you use depends to some extent on your own preferences. Follow these guidelines:

- **Resource Names column:** By default, you assign resources at 100 percent availability. If you want to assign a different percentage, don't use this method; it's more difficult if you need to assign multiple resources.
- **Assign Resources dialog box:** Replace one resource with another (using the handy Replace feature) or filter the list of available resources by a criterion (for example, resources with a cost of less than a specified amount). This method is useful for assigning multiple resources.
- **Task Information dialog box:** Having the task details handy (such as task type or the constraints contained on other tabs of this dialog box) is helpful when you make the assignment.

## Selecting resources from the Resource column

You can add resources from the Resource column, whether it's from Gantt Chart view or Tracking Gantt view.

WARNING

Even though Task Usage view lists tasks in its sheet pane and can even display a Resource column, this view can't be used to add resource assignments.

Follow these steps to assign resources at a default percentage:

1. **Display Gantt Chart view by clicking the Gantt Chart button on the Task tab or View tab of the Ribbon.**
2. **Click the View tab. In the Data group, click Tables. From the drop-down menu, make sure you are in the Entry table (Entry is the default).**
3. **Click in the Resource Names column for the task to which you want to make a resource assignment.**

   An arrow appears at the end of the cell.
4. **Click the arrow to display a list of resources.**

   Figure 9-2 shows a list of resources.
5. **Click the resource you want to assign.**

   The resource name appears in the Resource column, assigned at 100 percent.

TIP

You can always change the assignment units later by opening the Task Information dialog box and changing the assignment units on the Resources tab.

## Using the Assign Resources dialog box

To assign a work resource or material resource to a task, you can select a task and then use the Assign Resources dialog box to make assignments. This unique dialog box is the only one you can leave open and continue to scroll within the project, making it easy to switch between selecting a task and making an assignment. This method is easy to use for assigning material and cost resources. To use the Assign Resources dialog box, follow these steps:

1. **Click the Assign Resources button in the Assignments group of the Resource tab on the Ribbon.**

   The Assign Resources dialog box opens, as shown in Figure 9-3.
2. **Click a task to select it.**

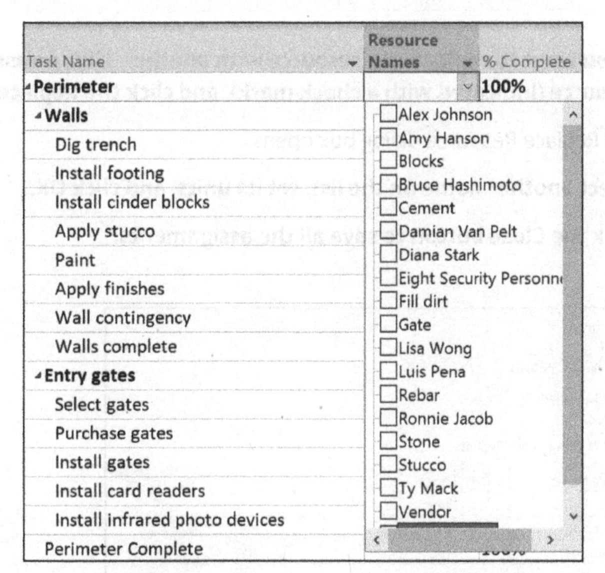


**FIGURE 9-2:** Resource list.

© John Wiley & Sons, Inc.

3. **Select the resource(s).**

   To select more than one resource, hold down the Control/Ctrl key and select as many resources as you need.

4. **Click the Units column for the work or material resource you want to assign and then specify the assignment units for the resource.**

   Click the spinner arrows in the Units box to increase or decrease the units. If you don't select units for a work resource, Project assumes that you're assigning 100 percent of the resource. The spinner arrows show 0 percent, 50 percent, or 100 percent, or you can simply type a percentage. For a material resource, use the spinner arrows in the Units column to increase or decrease the unit assignment, or type a number of units. For a cost resource, type the expected cost for that particular assignment.

5. **Click the Assign button.**

   The Assign Resources dialog box, with the task being assigned, appears. It is shown in Figure 9-3.

   The Assign Resources dialog box can remain open while you select tasks in the task list. Then you can make assignments to a variety of tasks without the need to close and reopen the dialog box repeatedly.

6. **Repeat Steps 2 through 5 to add all resources.**

7. **If you want to replace one resource with another, click an assigned resource (indicated with a check mark), and click the Replace button.**

   The Replace Resource dialog box opens.

8. **Select another name on the list, set its units, and click OK.**
9. **Click the Close button to save all the assignments.**

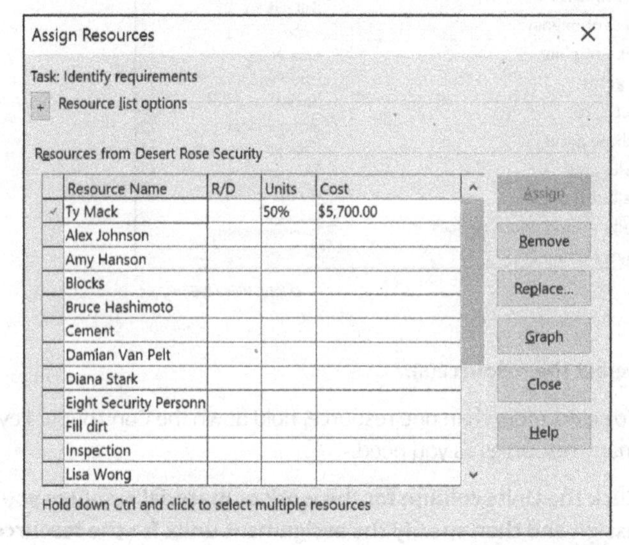


**FIGURE 9-3:** The Assign Resources dialog box.

*© John Wiley & Sons, Inc.*

## Adding assignments in the Task Information dialog box

You can assign resources on the Resources tab of the Task Information dialog box for any task by following these steps:

1. **Double-click a task name in Gantt Chart view.**

   The Task Information dialog box appears.

2. **Click the Resources tab to display it.**
3. **Click in a blank Resource Name box and then click the arrow that appears on the right side of the box.**

   A drop-down list of resources appears.

4. **Click the resource you want to assign.**

5. **For a work or material resource, click the Units column and use the spinner arrows to set an assignment percentage.**

   If you're assigning a material resource, the Units default setting is a single unit. (If your units are pounds, the default assignment is 1 lb.) Use the spinner arrows in the Unit field or type a value there to assign additional material units.

6. **Repeat Steps 3 through 5 to assign additional resources.**
7. **Click the OK button.**

## Shaping the contour that's right for you

When you make a work resource assignment, Project spreads out the work evenly over the life of the task. However, you can modify the level of work that takes place during the life of an autoscheduled task — its *work contour* — so that more work takes place near the beginning, middle, or end of the task.

For example, if you know that the employees on a task to install a new computer network will have to spend some time up front studying the manuals and reviewing the schematics for the wiring before they can begin to make measurable progress on the installation, you may use a late-peaking contour. Or if you know that employees are likely to put in a lot of work up front on a survey — and then sit back and wait for the results to come in — you may choose an early-peaking contour.

Using a different contour on a particular resource's task assignment can free that resource to work on a second task that occurs during the life of the first task. This strategy can help you resolve a resource conflict.

WARNING

The contour you select will have slightly different effects, depending on the task type. Trust me: Most project managers don't even want to try to understand this complex equation. Simply try a different contour to see whether it solves your problem and doesn't make too dramatic a change to the task duration or another resource assignment. Realize that most all the applied work contours result in a greater task duration.

To set a task's contour, follow these steps:

1. **On the View tab, in the Task Views group, select Task Usage.**

   This view shows resource assignments by task.

2. **Double-click a resource.**

   The Assignment Information dialog box appears.

3. **On the General tab, from the Work Contour drop-down menu, choose a preset pattern.**

   The Work Contour options are shown in Figure 9-4.

4. **Click the OK button to save the setting.**

TIP

If none of these patterns fits your situation, you can manually modify a resource's work by changing the number of hours the resource puts in day by day on a task in Task Usage view.

Assignment Information

General | Tracking | Notes

Task: Purchase Trucks

Resource: Vijay Chandra

Work: 60h Units: 50%

Work contour: Front Loaded

Flat
Back Loaded
Front Loaded
Double Peak
Early Peak
Late Peak
Bell
Turtle

Start: Booking type: Committed

Finish: Cost: $4,500.00

Cost rate table: Assignment Owner:

OK Cancel

**FIGURE 9-4:** Work Contour options.

© John Wiley & Sons, Inc.

**Before you save, ensure that your modifications equal the number of hours you want, or else you can inadvertently change the resource's assignment.**

**A Work Contour icon appears in the i column in Task Usage view, so you can easily see and identify the work contours you have set.**

TIP

**Resource Usage view is useful for reviewing resource hours across a timeline at this stage of planning. This view gives a side-by-side comparison between a single resource's workload and all the tasks taking place during a particular period in the project.**

# Benefitting from a Helpful Planner

One thing you may notice after making resource assignments is that a little Person icon appears as an indicator beside certain tasks in Gantt Chart view. If you switch to a resource-oriented view, you may see that certain resource names now display in red. These changes are Project's way of indicating resources in danger of being overbooked in the plan.

Team Planner view can help you ensure that you've distributed work among resources in a manageable way. To change to Team Planner view, click the Resource tab and click the Team Planner button on the far left side. This view enables you to move assignments among resources, assign tasks that haven't been assigned, and even schedule tasks that haven't been scheduled. Figure 9-5 shows Team Planner view.

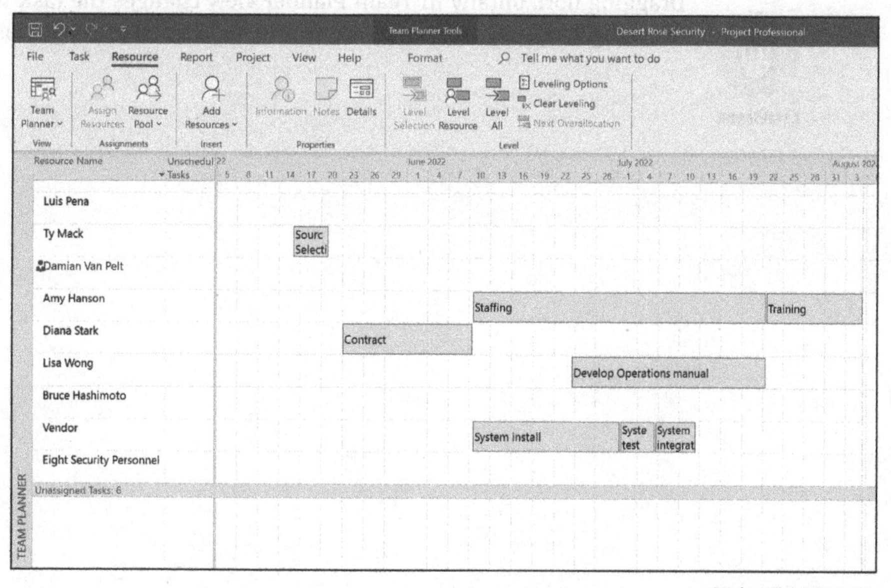


**FIGURE 9-5:** Team Planner view.

© John Wiley & Sons, Inc.

When two or more tasks assigned to a resource overlap and cause the resource to be overbooked (or *overallocated*), thin red bars appear above and below the overbooked period. Team Planner view also shows you any tasks that are unassigned or lack enough schedule information to be scheduled.

To view unassigned tasks, drag the bar at the bottom that shows the number of unassigned tasks up, dropping it to split the view the way you want.

From there, you can drag and drop to correct a number of schedule issues, and you can use the following techniques in any combination:

- **To fix a resource overallocation:** Drag an overlapping task to the left or right to change its schedule. If you want to reassign the task to another resource, drag the task up or down to that resource.
- **To assign an unassigned task:** Drag the task up to the row for the resource to which you want to assign the task.
- **To schedule an unscheduled task:** Drag the task to the right, to the timescale portion of the view, and then drop the task into a schedule position.

TIP

Team Planner view is only one of the tools you can use to handle overbooked resources. Chapter 12 presents more methods for fixing workload issues.

REMEMBER

Dragging horizontally in Team Planner view changes the task schedule, whereas dragging vertically from one resource to another changes the task assignment.

IN THIS CHAPTER

» Understanding how costs accrue

» Establishing work resource rates

» Specifying unit costs

» Adding fixed costs

» Allowing for overtime

» Adding up all the costs

# Chapter 10

# Determining a Project's Cost

There's no such thing as a free lunch — and if you use Project to track costs, there's no such thing as a free resource, because Project uses resources working on tasks as a way of calculating most of the costs of the project.

When you set up a resource, you specify a work resource rate (by default, this rate is calculated by the hour) or a material resource per use cost. You can also create *cost resources* — a variable cost that isn't calculated using a per-use or hourly rate but that may be used several times in a project, such as travel expenses. For example, if you assign a resource to work ten hours on a task and give that resource an hourly rate of $50, you've added a $500 cost to the project. Create a resource named *cement*, give it a unit cost of $200, and assign ten units (for example, ten tons of cement), and you've added a whopping $2,000 to the bottom line.

Some other factors come into play as well, such as how many hours a day a resource is available to work and any overtime rates. At the end of the day, all these settings come together to comprise the project cost.

This chapter helps you explore the relationship between resources and costs, and shows you how to set resource standard and overtime rates and create fixed costs.

# How Do Costs Accrue?

Project helps to account for costs on various tasks with a combination of costs per hour, costs per use, costs per unit, fixed costs, and costs for specific assignments of cost resources. Before you begin to flesh out cost information about resources, you have to understand how these calculations work.

Project provides two main pictures of the budget in a project: one at the moment you freeze the original plan (a *baseline plan*) and the ongoing picture of actual costs. The actual costs are determined from the activity and material usage you record as the project moves along. You record work effort on tasks, which have costs assigned to them. These costs then add up based on the effort expended or the units of materials used.

## Adding up the costs

The best way to understand how costs add up in the project is to look at an example. John Smith (not his real name) is managing a project that involves the building of a new gourmet ice-cream packaging plant. Here are the costs that John anticipates for the task he created, Install Ice Cream Mixers:

- About ten person-hours of effort to complete the installation
- A cost per use of $500 paid to the mixer manufacturer to oversee the installation and to train workers to use the machine
- Twenty pounds of ice cream ingredients to test the mixers
- Shipping expenses of $150 for the mixers and supplies
- A fixed cost of $2,500 for the mixers themselves

The ten hours of effort will be expended by work resources. The total cost for the ten hours is a calculation: 10 × the resource rate. If the resource rate is $20, this cost totals $200. If two resources work on the task, one at a rate of $20 and one at a rate of $30, then (with effort-driven scheduling turned off by default) Project adds 20 hours of effort between them, and the resulting cost is $500.

The cost per use of $500 is a fee to the manufacturer, which is also created as a work resource. This cost doesn't change based on the number of resources or the time involved.

The cost of 20 pounds of ice cream (any flavor you like) is calculated as 20 × the unit cost of the ice cream ingredients. If the unit cost is $2, this cost is $40.

Shipping the new equipment and ice cream supplies to you results in a charge of $150. Shipping was created as a cost resource because it has no associated work, and different shipping costs may apply to other tasks in the project.

Finally, the $2,500 cost for the mixers themselves was entered separately, as a fixed cost in the Cost table of Gantt Chart view.

And that's how costs are assigned and how they add up on projects. When working in Project, you soon discover that although it has quite capable costing features, they're probably some of the most underutilized.

TIP

You can create and assign resources that have no associated costs. For example, if you want your boss to be available to review status reports but your company doesn't require that your boss's time be charged to the project, you can simply use those resource assignments to remind you about the need for your boss's availability on that day or at that time.

## When will these costs hit the bottom line?

In business, you rarely get to choose when you pay your own bills. In Project, however, you can choose when your costs affect your budget.

Resources can be set to accrue at the start or end of the task that they're associated with or to be prorated throughout the life of the task. If a task of three months begins April 1, a $90 cost resource can be added to the actual costs to-date on Day 1, on Day 90, or at a dollar per day until the end of the task. Use the Accrue At field in Resource Sheet view or on the Cost Accrual drop-down list on the Costs tab of the Resource Information dialog box (shown in Figure 10-1) to set the accrual method.

This scenario isn't exactly a purely realistic reflection of how you have to pay for costs because, let's face it: Most bills come due 30 days after they hit your desk. It's more a factor of when you want that cost to show up for the purposes of tracking costs and reporting expenses against the project. My preference is to accrue cost at the same rate at which work is performed. This keeps costs and work aligned.

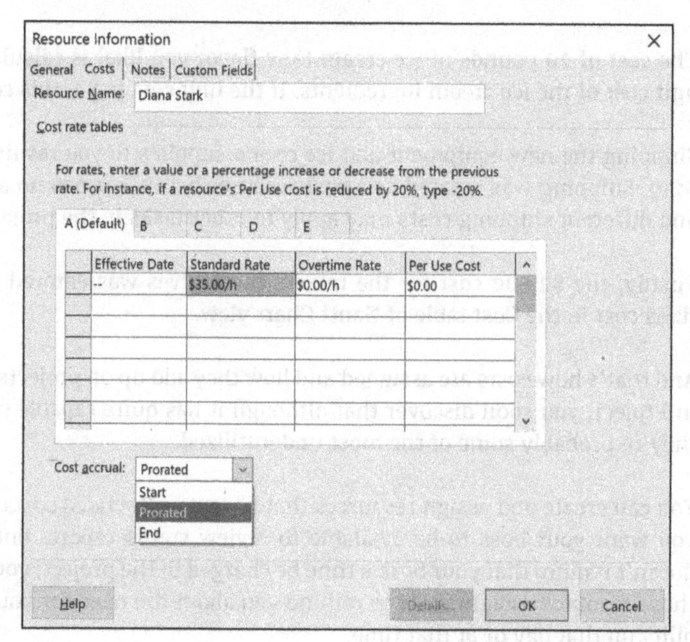


**FIGURE 10-1:** Cost accrual options in the Resource Information dialog box.



# Specifying Cost Information in the Project

Most projects involve a combination of cost types: work, material, and cost (that is, fixed cost). Before you can enter the information at the task or resource level, you have to do your homework to find out the fixed costs as well as the hourly or unit rates for all your resources.

During the planning stages, you may not be able to anticipate exactly what a particular cost will be or know every resource's rates. In this instance, you may need to develop an estimate and progressively elaborate the estimate as you know more. For example, if you know that you have to create an Operations manual, you might start with a high-level estimate. As you find out more about the content, the level of detail, and the resource skills you need, you can define costs at a more detailed and more accurate level. By starting with an estimate, at least some cost will be reflected in the plan, and you can enter more accurate information later, as soon as you know it.

TIP

Use a field in the Resource sheet, such as the Code field, to designate resources as having estimated rates or costs so that you can easily go back to those tasks and update the estimates as the plan progresses.

## You can't avoid fixed costs

Maybe it's that huge fee for the consulting company your boss insisted you use, even though you knew that the report wouldn't tell you a thing you didn't already know. Or perhaps it's the $2,000 for a laptop computer you talked your boss into buying so that you could manage the project when you're on the road. Whatever it is, it's a cost that doesn't change, no matter how many hours the task requires or how many people work on the task. It has no unit cost or rate per hour, and no actual hours of work are tracked. It's a *fixed cost.*

You can specify this type of cost by entering **Cost** in the Type field of the Resource sheet. However, you can't enter the information in the Standard Rate or Cost/Use columns in the Resource sheet. You need to specify the amount in the Cost column of the Resources tab, found on the Task Information dialog box when you add the resource to the task. Figure 10-2 shows the Task Information box with information in the Resources tab. Note that team resource's cost is calculated automatically based on information entered in the Resource sheet. However, the cost for the Gate was entered manually. Every time you assign a Cost resource to a task, you have to specify the actual cost associated with it.

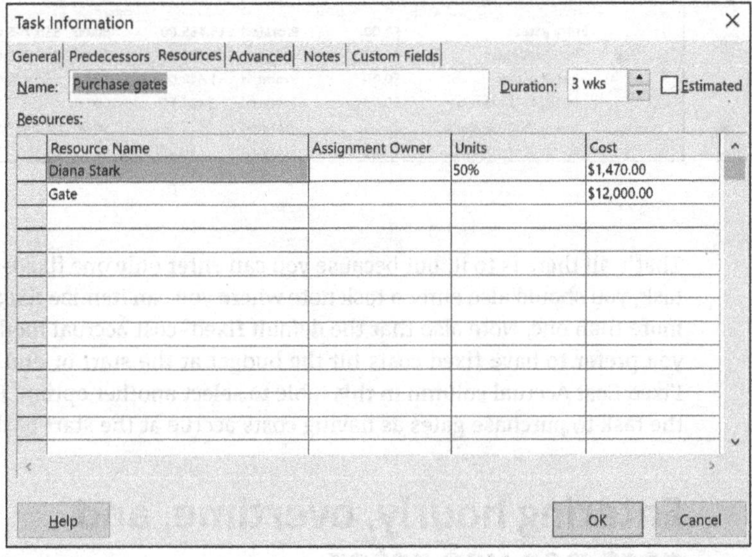


**FIGURE 10-2:** Cost information in the Task Information dialog box.

© John Wiley & Sons, Inc.

You can also simply enter a fixed cost associated with a task without having to create and assign a cost resource to it. To do so, you can use the Cost table in the sheet in Gantt Chart view.

REMEMBER

A *table* is a preset column combination that simplifies entering certain information in a sheet pane.

Follow these steps to enter a fixed cost for a task:

1. **Display the project in Gantt Chart view.**
2. **Click the View tab, and in the Data group, choose Tables ⇨ Cost.**

   The table of columns appears, as shown in Figure 10-3. You can insert the Fixed Cost column into any sheet, but the Cost table is ready for you to use.

3. **Click the Fixed Cost column for the task to which you want to assign the cost and then enter the amount.**

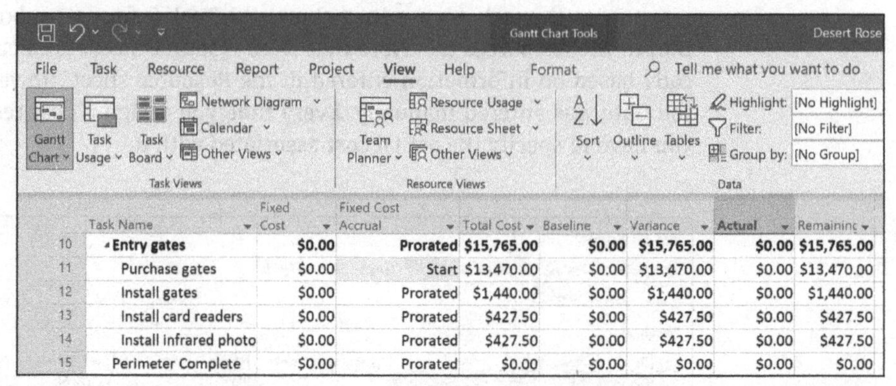


| | Task Name | Fixed Cost | Fixed Cost Accrual | Total Cost | Baseline | Variance | Actual | Remaining |
|---|---|---|---|---|---|---|---|---|
| 10 | Entry gates | $0.00 | Prorated | $15,765.00 | $0.00 | $15,765.00 | $0.00 | $15,765.00 |
| 11 | Purchase gates | $0.00 | Start | $13,470.00 | $0.00 | $13,470.00 | $0.00 | $13,470.00 |
| 12 | Install gates | $0.00 | Prorated | $1,440.00 | $0.00 | $1,440.00 | $0.00 | $1,440.00 |
| 13 | Install card readers | $0.00 | Prorated | $427.50 | $0.00 | $427.50 | $0.00 | $427.50 |
| 14 | Install infrared photo | $0.00 | Prorated | $427.50 | $0.00 | $427.50 | $0.00 | $427.50 |
| 15 | Perimeter Complete | $0.00 | Prorated | $0.00 | $0.00 | $0.00 | $0.00 | $0.00 |

**FIGURE 10-3:** The Cost table.



That's all there is to it, but because you can enter only one fixed-cost amount for a task, you should also enter a task note where you can itemize fixed costs if you have more than one. Note also that the default fixed-cost accrual method is prorated: If you prefer to have fixed costs hit the budget at the start or end of a task, use the Fixed Cost Accrual column in this table to select another option. Figure 10-3 shows the task to purchase gates as having costs accrue at the start of the task.

## Entering hourly, overtime, and cost-per-use rates

Whether it's minimum wage or the astronomical fees your lawyer charges you every time you sneeze, most people are paid a certain amount per hour. To represent most people involved in the project, you create work resources and charge them to the project at an hourly rate. Some resources also charge an additional flat fee for each use. For example, a plumber may charge a trip charge or "show up" fee of $75 and then add the cost per hour. This flat fee is a cost per use, added every time you assign the resource to a task, no matter how many hours are involved in completing the task.

The first cost you enter is a resource's hourly rate and an overtime rate (if the resource has one). After a resource has an hourly rate, you can enter the estimate of how many hours that person will work on each task they are assigned to, and Project totals their estimated costs in your plan and adds any cost per use. When you track actual effort expended on tasks, the actual costs are determined by a calculation of this actual effort multiplied by the hourly rate, plus any cost per use.

REMEMBER

By comparing estimated costs and actual costs, you form an ongoing picture of whether the project is within budget.

To set resource rates per hour and cost per use for a work resource, follow these steps:

1. **Display Resource Sheet view.**
2. **Click the Std Rate column for the resource to which you want to assign a cost.**
3. **Enter a dollar amount.**

   If you're entering a rate for a unit other than hours, type a slash (/) and then the unit (for example, y for year or mo for month).
4. **Press Tab to go to the Ovt column.**
5. **Enter a dollar amount or rate again.**
6. **Press Tab to move to the Cost/Use column.**
7. **Type a dollar amount.**

   This amount is the flat-fee amount that will be added every time you assign the resource to a task.
8. **Press Tab.**

   The entry is completed.

Figure 10-4 shows the work resources with their standard hourly rates and some with overtime rates.

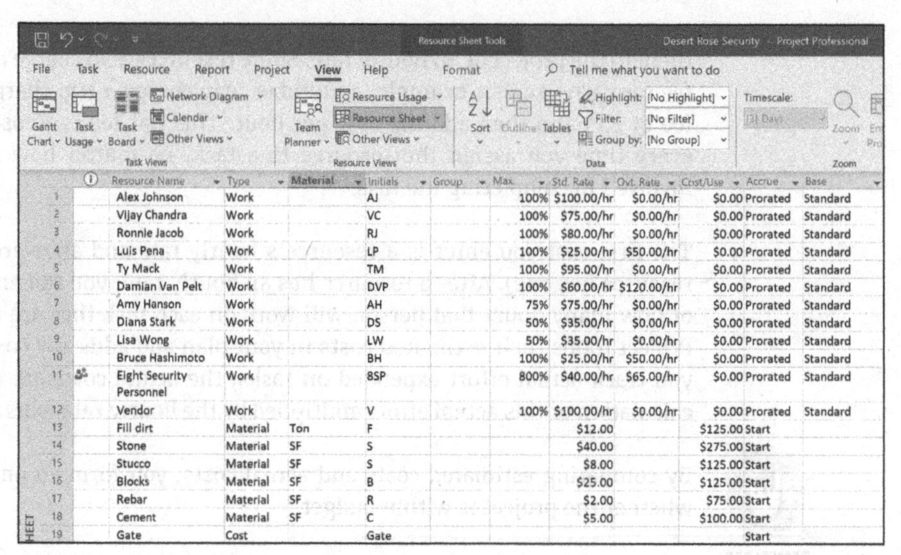

| | | Resource Name | Type | Material | Initials | Group | Max. | Std. Rate | Ovt. Rate | Cost/Use | Accrue | Base |
|---|---|---|---|---|---|---|---|---|---|---|---|---|
| 1 | | Alex Johnson | Work | | AJ | | 100% | $100.00/hr | $0.00/hr | $0.00 | Prorated | Standard |
| 2 | | Vijay Chandra | Work | | VC | | 100% | $75.00/hr | $0.00/hr | $0.00 | Prorated | Standard |
| 3 | | Ronnie Jacob | Work | | RJ | | 100% | $80.00/hr | $0.00/hr | $0.00 | Prorated | Standard |
| 4 | | Luis Pena | Work | | LP | | 100% | $25.00/hr | $50.00/hr | $0.00 | Prorated | Standard |
| 5 | | Ty Mack | Work | | TM | | 100% | $95.00/hr | $0.00/hr | $0.00 | Prorated | Standard |
| 6 | | Damian Van Pelt | Work | | DVP | | 100% | $60.00/hr | $120.00/hr | $0.00 | Prorated | Standard |
| 7 | | Amy Hanson | Work | | AH | | 75% | $75.00/hr | $0.00/hr | $0.00 | Prorated | Standard |
| 8 | | Diana Stark | Work | | DS | | 50% | $35.00/hr | $0.00/hr | $0.00 | Prorated | Standard |
| 9 | | Lisa Wong | Work | | LW | | 50% | $35.00/hr | $0.00/hr | $0.00 | Prorated | Standard |
| 10 | | Bruce Hashimoto | Work | | BH | | 100% | $25.00/hr | $50.00/hr | $0.00 | Prorated | Standard |
| 11 | | Eight Security Personnel | Work | | 8SP | | 800% | $40.00/hr | $65.00/hr | $0.00 | Prorated | Standard |
| 12 | | Vendor | Work | | V | | 100% | $100.00/hr | $0.00/hr | $0.00 | Prorated | Standard |
| 13 | | Fill dirt | Material | Ton | F | | | $12.00 | | $125.00 | Start | |
| 14 | | Stone | Material | SF | S | | | $40.00 | | $275.00 | Start | |
| 15 | | Stucco | Material | SF | S | | | $8.00 | | $125.00 | Start | |
| 16 | | Blocks | Material | SF | B | | | $25.00 | | $125.00 | Start | |
| 17 | | Rebar | Material | SF | R | | | $2.00 | | $75.00 | Start | |
| 18 | | Cement | Material | SF | C | | | $5.00 | | $100.00 | Start | |
| 19 | | Gate | Cost | | Gate | | | | | | Start | |

**FIGURE 10-4:** Resource costs.

TIP

In many projects, you don't track the cost per resource for internal resources because people's salaried rates are generally considered confidential. Sometimes, a standard midrange or loaded-resource rate is used to track employee rates by job grade.

## Assigning material resources

Calculating the cost of a material resource might take you back to solving problems in your old high school algebra class. Fortunately, if you aren't an algebra whiz, Project makes a straightforward calculation to arrive at the cost of using a material resource.

When you assign a work resource to a task, Project multiplies its standard hourly rate by the hours of work for the assignment. But material resources don't have hours of work: You pay for them by the unit quantity, not by the hour. So when you set up a material resource, you specify a standard rate for a single unit (per yard, or ton, or gallon, for example) and assign a certain number of units to each task. The cost is the number of units multiplied by the cost per use.

To assign a standard unit rate for a material resource, follow these steps:

1. **Display Resource Sheet view.**
2. **If you haven't already done so, click the Material column for that resource and then type a unit name (such as gallon).**
3. **Click the Std Rate column for the resource you want to set and then type a dollar amount (such as the cost per gallon).**
4. **Press Enter or Tab to finish the entry.**

Figure 10-4 shows that fill dirt is set up as $12 per ton, with a delivery fee in the Cost/Use column of $125. Other material resources are calculated per square foot (SF).

You can also make cost entries in the Costs tab of the Resource Information dialog box. Figure 10-5 shows the columns labeled Standard Rate, Overtime Rate, and Per Use Cost.

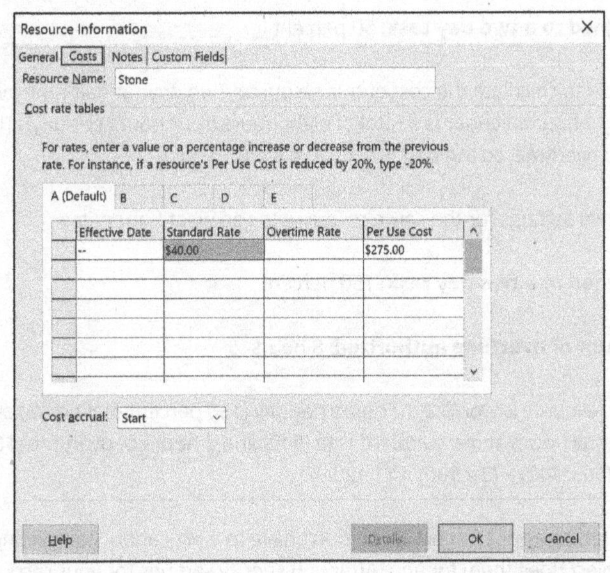


**FIGURE 10-5:** Setting rates in the Resource Information dialog box.

© John Wiley & Sons, Inc.

Note that you can also use the Resource Information dialog box to enter as many as five standard unit rates with effective dates to account for fluctuations in unit cost over the life of the project.

## HOW YOUR SETTINGS AFFECT YOUR COSTS

In Chapter 9, I describe how to assign resources to tasks and how to assign those resources in certain units (percentages for work resources and quantity consumed for material resources). All these factors work together in calculating the cost of the resource when assigned to tasks.

Suppose that you want to assign a mechanic to a task with the following parameters:

**Base calendar:** Night Shift (eight hours, six days per week, between 11 p.m. and 7 a.m.)

**Cost per hour:** $40

**Overtime cost:** $60

**Availability:** 100 percent

**Assigned to a two-day task:** 50 percent

Here's how to calculate the cost of this resource: Two days at half-time availability based on an eight-hour calendar is a total of eight hours (four hours per day). The resource incurs no overtime, so the cost is 8 × $40 = $320.

Change two settings for the same resource to see what happens:

**Assigned to a two-day task:** 150 percent

**Amount of overtime authorized:** 8 hours

Now the resource is working 12 hours per day (150 percent of 8 hours) over two days. With 16 total hours at the standard rate ($40) and 8 hours of overtime ($60), this person will cost (16 × $40) + (8 × $60) = $1,120.

The great thing about Project is you don't have to worry about completing the calculations; Project does them for you. After you specify settings for your resources, Project does the work of tallying and showing total costs to you.

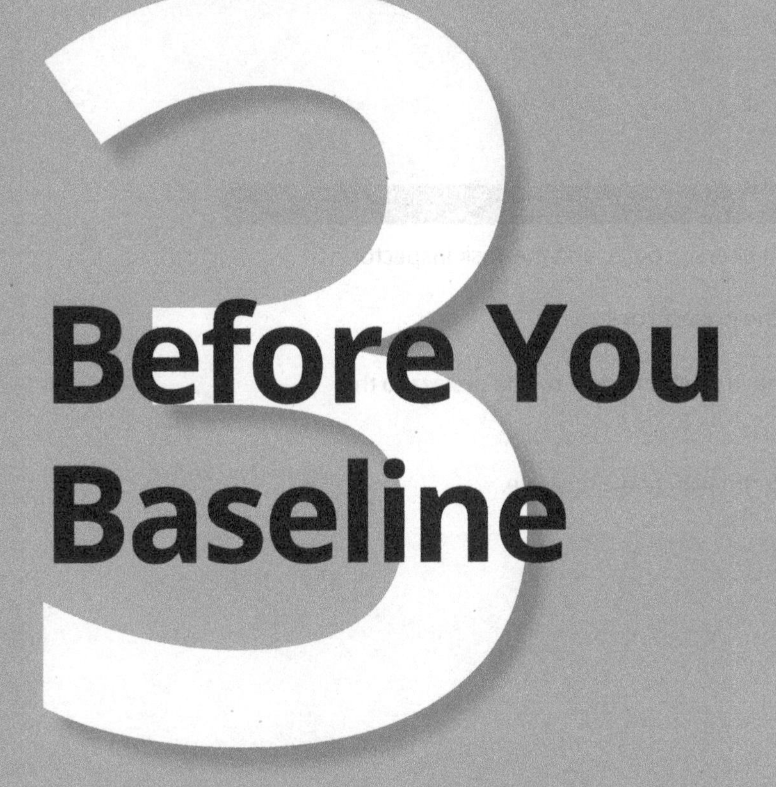

# Before You Baseline

IN THIS PART . . .

Work with filters, groups, and the Task Inspector.

Manage the project constraints.

Format the Project schedule to add pizzazz to the project.

Set an interim plan and a baseline.

IN THIS CHAPTER

» Viewing timing and resource issues with filters

» Grouping information

» Using the Task Inspector pane to determine task drivers

» Working with warnings, suggestions, and problems to fix schedules

# Chapter 11 Fine-Tuning Your Plan

As they say, the best-laid schemes of mice and project managers go oft awry, and the schedule is no different. After you take your best shot at laying out the project schedule, creating every task, and assigning every resource — and you think you're ready to start your project — think again.

A close look at almost any schedule reveals issues that you should resolve before you baseline. These issues may include a schedule that ends a month after the projected finish date, human resources who are assigned to work 36-hour days, or a budget that exceeds the national debt. (Details, details. . . .)

Even if you see no glaring problems related to time, workload, or cost, you should ensure that the project schedule is as realistic as possible before you commit to it. Take a moment to give the project the once-over by using filters to focus your attention on potential problem areas. You can also use the Task Inspector to pinpoint issues with the schedule or resources. If these options aren't enough, Project identifies problem areas and suggests solutions.

## Everything Filters to the Bottom Line

The first step in solidifying the project schedule is simply to look at it from a few different perspectives — similar to circling a car to inspect all its features before you fork over the down payment. Filters help you gain that kind of perspective.

At this stage, you can use filters to examine two major problem areas:

- **Overallocated resources:** Those resources working more than the number of hours you specified.
- **Tasks on a critical path:** The *critical path* consists of the series of tasks in the project that must be completed on time for the project to meet its final finish date.

One way to think about the critical path is that it's the one with the least amount of slack (or *float*). Any task that has *slack* — that is, any length of time that the task could be delayed without affecting a milestone or the project end date — is not on the critical path. If the project has multiple critical paths, or near-critical paths, delays on those paths are likely to cause a problem as well. You can say that for every day the critical path slips, the project delivery date slips a day. Therefore, you can see the necessity of not only knowing the project's critical (and near-critical) paths, but also being able to easily check their status.

## Setting predesigned filters

Filters give you a close-up look at various aspects of your plan and help you spot clues about problems, such as overallocated resources. You can set a filter to highlight tasks or resources that meet certain criteria or to remove from view any tasks or resources that don't meet the criteria.

Project provides predesigned filters that you can simply turn on for tasks or resources, using criteria such as:

- Tasks with a cost more than a specified amount
- Tasks on the critical path
- Tasks that occur within a certain date range
- Milestone tasks
- Tasks that use resources in a resource group
- Tasks with overallocated resources

TIP

Several filters, such as Slipping Tasks and Over-Budget Work, help you spot problems after you've finalized the schedule and are tracking progress. (See Chapter 15 for more about tracking.)

You can access filters in a couple of ways. When you use the Filter drop-down list, you choose built-in filters from a list. The filters act to remove from view any tasks that don't meet specified criteria and display the ones that do.

To turn on these filters from the View tab of the Ribbon, follow these steps:

1. **Display a resource view (such as Resource Sheet view) to filter for resources or a task view (such as Gantt Chart view) to filter for tasks.**
2. **In the Data group on the View tab, click the Filter drop-down list and choose a criterion.**

   The Filter list is a drop-down list. If you choose a filter that requires input, you see a dialog box, such as the Resource Range one shown in Figure 11-1. The dialog box will be different depending on which filter you choose. If the filter requires no input, the filter is applied immediately, and it removes from view any resources or tasks that don't match your criteria.
3. **If a dialog box is displayed, fill in the required information and click the OK button.**

   The filter is applied.

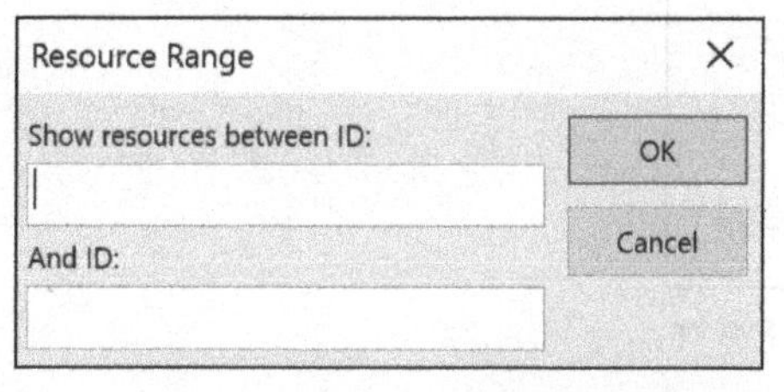


**FIGURE 11-1:** Entering filter parameters.

© John Wiley & Sons, Inc.

To redisplay all tasks or resources, click the Filter drop-down list on the View tab and then select the No Filter option.

## Putting AutoFilter to work

The Project AutoFilter feature is turned on by default for all new schedule files. Arrows appear in the column headings in the displayed sheet. When you click the arrow in the Resource Names column, for example, the name of every resource assigned to tasks in the project is listed in alphabetical order, along with other filtering and sorting choices. See Figure 11-2 for an example. In the AutoFilter list for a column, select check boxes to control which items appear onscreen. Selected items appear, and deselected items are hidden by the filter. If you want to display only a few items, clear the Select All check box first to deselect all items and then select the check boxes next to individual items to reselect them. Then click OK and your filter is applied.

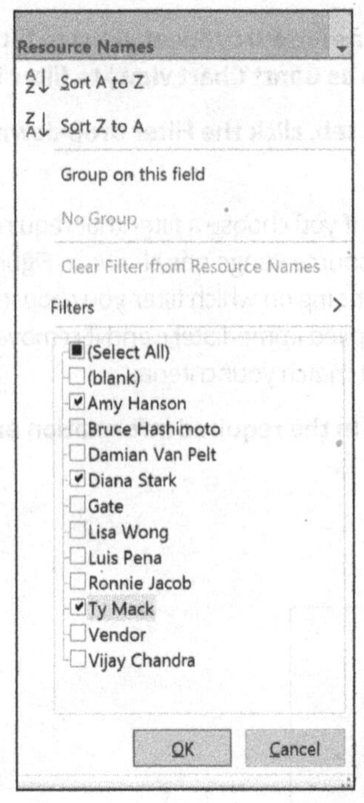


**FIGURE 11-2:** Using check boxes with AutoFilter.



Follow these steps to use AutoFilter:

1. **In the View tab, go to the Data group and click the down arrow in the filter tool.**
2. **Use the Filter drop-down list to select a filter.**

   Select a predefined filter, as shown in Figure 11-3. For example, if you're filtering for incomplete tasks, you see any tasks that do not show 100 percent complete. The Clear Filter choice removes a previously applied filter.
3. **Click the OK button.**

   Any task that doesn't meet your criteria temporarily disappears from the view.

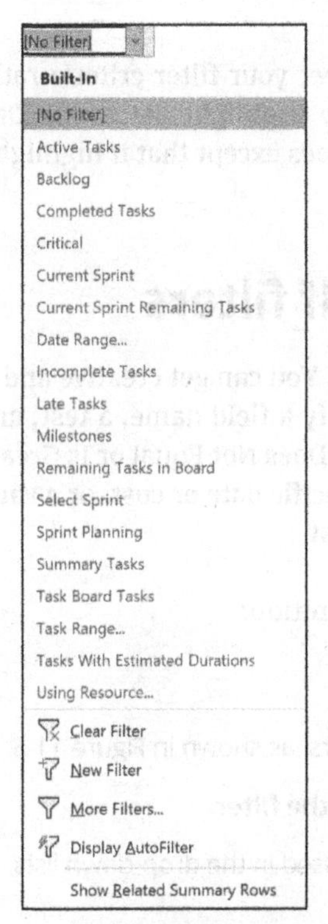


**FIGURE 11-3:** The Filter drop-down list.

© John Wiley & Sons, Inc.

If you don't see a filter you want, you can select More Filters in the drop-down list. The More Filters dialog box will appear, as shown in Figure 11-4. Choose the Task or Resource radio button to see all the available filters, including ones you create.

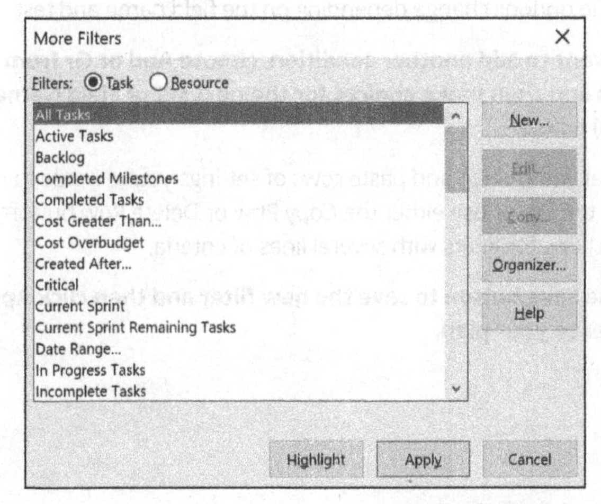


**FIGURE 11-4:** The More Filters dialog box.

© John Wiley & Sons, Inc.

TIP

To apply highlighting to items that meet your filter criteria rather than remove nonmatching items from view, use the Highlight list in the Data group on the View tab. This list works like the filter does except that it highlights the applicable tasks instead of filtering tasks out.

## Creating do-it-yourself filters

You don't have to use predefined filters: You can get creative and design your own filters. To define a new filter, you specify a field name, a test, and a value. A test is a condition that must be met, such as Does Not Equal or Is Greater Than. A value is either a value you enter, such as a specific date or cost, or a value you pick from the drop-down list, such as baseline cost.

Here's how to build your own filter definition:

1. **Choose View ➪ Filter ➪ New Filter.**

   The Filter Definition dialog box appears, as shown in Figure 11-5.
2. **In the Name field, type a name for the filter.**

   If you want the filter to be easily accessed in the drop-down lists, select the Show in Menu check box.
3. **Click the first line of the Field Name column and then click the down arrow that appears to display the list of choices.**

   Click a field name to select it.
4. **Choose a test, such as Equals, Is Greater Than, Is Within, and so on.**
5. **Choose a value.**

   The Value options change depending on the field name and test.
6. **If you want to add another condition, choose And or Or from the And/Or column and then make choices for the next set of Field Name, Test, and Value(s) boxes.**

   Note that you can cut and paste rows of settings you've made to rearrange them in the list, or use either the Copy Row or Delete Row button to perform those actions for filters with several lines of criteria.
7. **Click the Save button to save the new filter and then click Apply to apply the filter to your plan.**

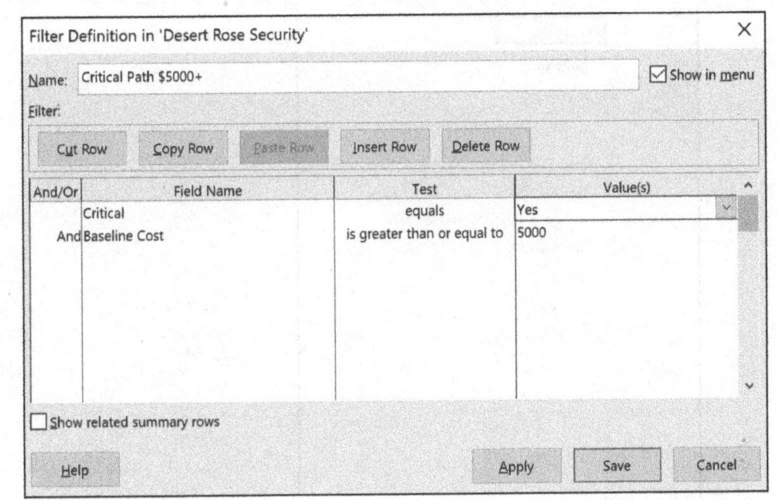


**FIGURE 11-5:** The Filter Definition dialog box.

*© John Wiley & Sons, Inc.*

TIP

Click the Organizer button in the More Filters dialog box to copy filters you've created from one Project file to another file or to make them available in all your files by copying the new filter into the `Global.mpt`.

# Gathering Information in Groups

Project uses the Group feature to chunk information into logical groupings based on predefined criteria. For example, you can use the Group feature if you want to see resources organized by work group, or you may organize tasks by their duration, from shortest to longest.

Organizing tasks or resources in this way may help you spot a potential problem in the project — for example, if you find that the majority of resources at the project startup are unskilled or that most of the tasks at the end of the project are on the critical path. Like filters, groups come predefined, as shown in Figure 11-6. If you don't see the group you like, click More Groups to display the More Groups dialog box (see Figure 11-7) or create a custom group. Groups also have a function similar to outlining, in which you can collapse and expand groups to better zero in on the tasks or resources that you want to examine more closely.

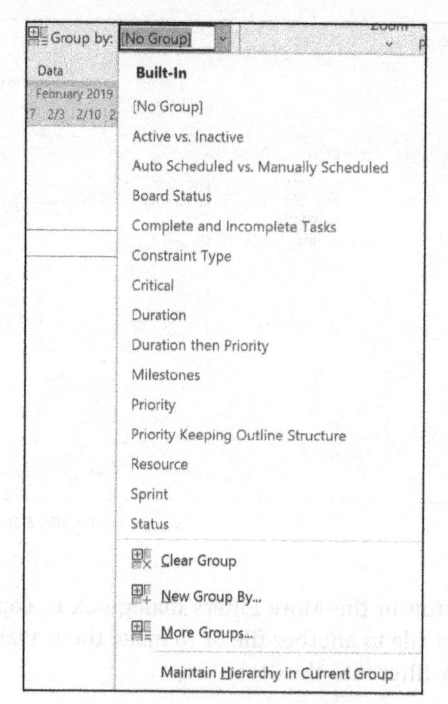


**FIGURE 11-6:** The Group drop-down list.

© John Wiley & Sons, Inc.

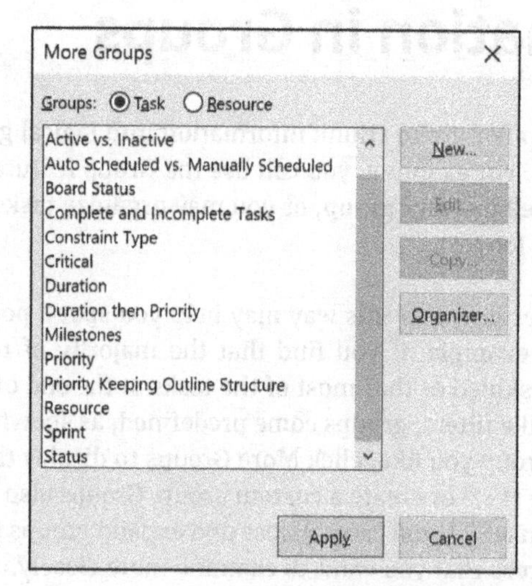


**FIGURE 11-7:** The More Groups dialog box.

© John Wiley & Sons, Inc.

## Applying predefined groups

Predefined groups, which are quick and easy to apply, cover a host of common requirements in projects. Follow these steps to apply a predefined group structure to the project:

1. **Display either a resource view (such as Resource Sheet view) to group resources or a task view (such as Gantt Chart view) to group tasks.**
2. **On the View tab, click the Group By drop-down button in the Data group and then choose a group.**

   The information is organized according to your selection. Figure 11-8 shows Gantt Chart view grouped by milestone. Tasks that aren't milestones are collapsed.

| Task Name | Duration | Predecessors |
|---|---|---|
| **Milestone: Yes** | **109.63d** | |
| Walls complete | 0 days | 8 |
| Perimeter Complete | 0 days | 13,14,15 |
| Vehicles Complete | 0 days | 19,20 |
| Communication Equipment Complete | 0 days | 23,24,25 |
| System complete | 0 days | 35 |
| Release | 0 days | 43 |
| Operations Readiness Complete | 0 days | 48 |
| **Milestone: No** | **153.63d** | |

**FIGURE 11-8:** Tasks grouped by milestones.



To redisplay all tasks or resources in their original order, click the arrow in the Group list on the View tab to display the list. Then select the No Group option. (When no group is applied, the Group box displays `[No Group]`.)

## Devising your own groups

Custom groups include three elements: a field name, a field type, and an order. For example, you may create a group that shows the field name (such as Baseline Work) and the field type (such as Task, Resource, or Assignment) in a certain order (descending or ascending). A group that shows Baseline Work for tasks in descending order, for example, would list tasks in order from the greatest number of work hours required to the least number required. Other settings you can specify for groups control the format of the group's appearance, such as its font and font color.

Follow these steps to create a custom group:

1. **Choose View ⇨ Group By ⇨ New Group By.**

   The Group Definition dialog box appears, as shown in Figure 11-9.

2. **In the Name field, type a name for the group.**

3. **Click the first line of the Field Name column, click the down arrow that appears to display the list of choices, and then click a field name to choose it.**

4. **Repeat Step 3 for the Field Type and Order columns.**

   Note that if you want the Field Type option of grouping by assignment rather than by resource or task, you must first check the Group Assignments, Not Tasks check box to make that field available to you. Otherwise, the field type of Task or Resource appears by default.

5. **If you want to add another sorting criterion, click a Then By row and make choices for the Field Name, Field Type, and Order columns.**

6. **If you want the new group to be shown in the list when you click the Group drop-down list in the Data group on the View tab, select the Show in Menu check box.**

7. **Depending on the field name you've chosen, you can make settings for the font, cell background, and pattern to format the group.**

   I tell you more about formatting the project schedule in Chapter 13.

8. **If you want to define intervals in which to organize the groups, click the Define Group Intervals button.**

   This step displays the Define Group Interval dialog box, as shown in Figure 11-10. Use these settings to specify an interval. For example, if the Group By criterion is Duration, you can set the group interval to anything from minutes all the way up to months.

9. **Click the Save button to save the new group and then click Apply in the More Groups dialog box to apply the group to your plan.**

TIP

If you want to make changes to an existing predefined group, choose View ⇨ Group By ⇨ More Groups. In the More Groups dialog box, make sure that the group is selected and then click the Edit button. The Group Definition dialog box then appears so that you can edit all those settings for an existing group.

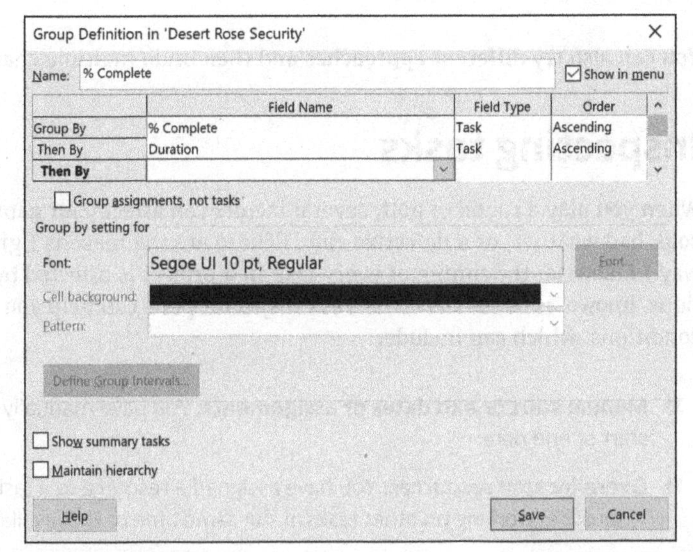


**FIGURE 11-9:** The Group Definition dialog box.

© John Wiley & Sons, Inc.

Define Group Interval
Field name: Duration
Group on: Each Value
Each Value
Start at:
Minutes
Hours
Group interval:
Days
Weeks
Months

**FIGURE 11-10:** The Define Group Interval dialog box.

© John Wiley & Sons, Inc.

# Figuring Out What's Driving the Project

With all the activity taking place in a project — perhaps hundreds of tasks, thousands of dependencies, and a busload of resources, for example — identifying the factors driving a project can be difficult. Two features in Project can help you see what's driving the project and make those all-important tweaks before finalizing the project plan:

- **Task Inspector pane:** Identifies the factors driving the timing of tasks in the project, pinpoints problems, and suggests solutions
- **Task warnings and suggestions:** Help you identify possible schedule issues in manually scheduled tasks

TIP

You can also try different approaches and then undo multiple changes.

## Inspecting tasks

When you play a round of golf, several factors can affect your game: an oncoming cold, bad weather, or a defective club. (These are the reasons I give myself, anyway.) Likewise, the timing of every task in a project is affected by certain conditions, known as *task drivers*. The Task Inspector pane can help you recognize these conditions, which can include:

- **Manual start or end dates or assignments:** You have manually entered a start or end date.
- **Overallocated resources:** You have assigned a resource to a task and the resource is working on other tasks at the same time or isn't available.
- **Calendars:** You have assigned a different calendar to the task or resource, or both, and the calendar differences are causing a scheduling conflict.
- **Leveling delay:** If you've turned on leveling to handle resource overallocation, the leveling may have caused a task to be delayed. (I discuss leveling in Chapter 12.)
- **Constraints:** You have applied a constraint to a task, such as forcing it to start or finish on a certain date.
- **Summary tasks:** You have manually scheduled a summary task that is out of sync with its child or subtasks.
- **Dependency relationships:** A predecessor task is causing issues with a task's timing.

Whether you use auto-scheduled tasks or manually scheduled tasks, you can use the Task Inspector to ensure the task schedule you've established works in conjunction with other tasks. Project performs behind-the-scenes calculations to determine when each manually scheduled task is likely to occur based on predecessors and its duration. If Project detects that the task is likely to get off track, it warns you about the potential problem and suggests a solution.

To display the Task Inspector pane, change to a task-oriented view such as Gantt Chart view and simply click the task you want to inspect, then click the Inspect button in the Tasks group on the Task tab. The Task Inspector pane appears. The Inspector pane explains the various conditions driving the timing of the selected task, as shown in Figure 11-11. If the Task Inspector pane includes a schedule change options, as shown in the figure, you can click the button for a solution to apply it to the task. Select another task to inspect it, or click the Close button in the Task Inspector pane when you're done.

Inspector

Purchase Phones
Resources overallocated due to work on other tasks:
Vijay Chandra

Move task to resource's next available time.

View overallocated resources in Team Planner.

Info

Factors affecting the task's start date:

Auto Scheduled

Constraint:
Type: Start No Earlier Than
Date: 3/23/22

Predecessors
19 - Purchase Trucks

GANTT CHART

| | Task Name | Duration | Predecessors | Resource Names |
|---|---|---|---|---|
| 18 | **Vehicles** | **20 days** | | |
| 19 | Purchase Trucks | 3 wks | 16 | Vijay Chandra[50%] |
| 20 | Purchase Carts | 4 wks | 19SS | Vijay Chandra[50%] |
| 21 | Vehicles Complete | 0 days | 19,20 | |
| 22 | **Communication** | **15 days** | | |
| 23 | Purchase Radios | 3 wks | 19SS | Vijay Chandra[50%] |
| 24 | Purchase Phones | 1 wk | 19SS | Vijay Chandra[50%] |
| 25 | Purchase Tablets | 1 wk | 19SS | Vijay Chandra[50%] |
| 26 | Communication Equipment Complete | 0 days | 23,24,25 | |
| 27 | **Security system** | **95.63 days** | | |
| 37 | **Asset Management** | **47 days** | | |
| 45 | **Operations Readiness** | **38.5 days** | | |

**FIGURE 11-11:** The Task Inspector pane.



## Handling task warnings, suggestions, and problems

Project performs forecasted calculations for each task, given the task drivers (dependencies, duration, constraints, calendars, and resource allocations). If Project finds a potential problem, it alerts you with an icon in the indicator column and in the Task Inspector.

You may see a red resource to indicate overallocated resources, a calendar to indicate a constraint, or a yellow triangle with an exclamation point inside it to indicate a warning. If the Task Inspector isn't open, you can hover over the icon in the indicator column to see the issues. By right-clicking the task, you can see your options.

Here's a rundown of the options you might see:

- **Reschedule to Available Date:** If an overallocated resource is assigned to the task, this choice reschedules all or part of the task to a time when the resource is available to handle the work.
- **Respect Links:** This warning solution moves the task based on the timing of its predecessor task. In most cases, this solution moves the task later based on the predecessor task's schedule.
- **Switch to Auto Scheduled:** This warning solution typically appears for manually scheduled tasks. Choosing this schedule change causes the tasks to recalculate based on the task drivers, such as dependencies and resource availability.

» **Ignore Problems for This Task:** This option allows the problem or overallocated resources to remain. It will keep the warning indicator if the task is still expected to go past its deadline.

Figure 11-12 shows several tasks with resource overallocation indicators and a few tasks with constraints. You can also see options for solving the issues when you right-click the task.

| | ⓘ | Task Name | Duration |
|---|---|---|---|
| 18 | | **Vehicles** | **20 days** |
| 19 | | Purchase Trucks | 3 wks |
| 20 | | Purchase Carts | 4 wks |
| 21 | | Vehicles Complete | 0 days |
| 22 | | **Communication** | **15 days** |
| 23 | | Purchase Radios | 3 wks |
| 24 | | Purchase Phones | 1 wk |
| 25 | | Purchase Tablets | 1 wk |
| 26 | | Communication Equipment Complete | 0 days |
| 27 | | **Security system** | **95.63 days** |
| 37 | | **Asset Manage** | |
| 45 | | **Operations Re** | |
| 46 | | Staffing | 6 wks |
| 47 | | Develop Oper | |
| 48 | | Training | |
| 49 | | Operations Re | |

Calibri 11

Fix in Task Inspector...

Reschedule to Available Date

Switch to Auto Scheduled

Ignore Problems for This Task

**FIGURE 11-12:** Tasks with icons.

© John Wiley & Sons, Inc.

TIP

You can click the drop-down arrow on the Inspect button in the Task group and select Show Warnings, Show Suggestions, and Show Ignored Problems. I usually only do this right before I baseline; otherwise, the warnings and suggestions become distracting.

IN THIS CHAPTER

» Adding contingency reserve to tasks

» Making adjustments to shorten the schedule

» Managing costs

» Resolving resource conflicts

» Moving a project to a better time

# Chapter 12 Negotiating Project Constraints

After you know which issues are affecting the project schedule, you can start to develop ways to address them. Sometimes, you simply have to recognize that events (especially on projects) don't always play out as planned. A sprinkling of contingency reserve may resolve the issue.

After you work out the initial version of the schedule, budget, and resource assignments, you meet with stakeholders to negotiate changes based on their priorities. For example, you might ask these questions:

- Can the schedule be altered to contain costs?
- Can in-house resources be added to expedite the work?
- Can less-expensive resources be hired to save money?

In this chapter, I describe multiple options for negotiating the various constraints of time, resources, scope, and costs that affect the project.

# It's about Time

Suppose that the boss asks you to commit to completing a project by a certain date. Your palms get sweaty and the pit of your stomach feels queasy, but then, for cover, you tack on a week to the targeted finish date and promise to deliver the impossible.

You hope that you can do it. You want to do it. But *can* you do it?

Project helps you feel confident about committing to a timeframe, because you can see how long every task takes to complete. Before you make promises to the boss, you had better know the total length of time to complete the project and its *critical path* — the longest series of tasks that must be completed on time in order to meet the overall project finish date.

The timing data for the project summary task tells you how long the entire project will take to complete. Simply display Gantt Chart view and look at the Duration, Start, and Finish columns. If the finish date isn't feasible, you have to modify tasks or resources or negotiate a different date.

Before you commit to an unreasonable date, remember Murphy's Law: Anything that can go wrong will go wrong. And Murphy *thrives* on projects. To lessen Murphy's influence, you allocate contingency reserve.

## Applying contingency reserve

*Contingency reserve* is either time or money that is factored into the schedule or the budget to respond to identified risks. A *risk* is an uncertain event or condition that, if it occurs, affects the schedule (or another project objective, such as cost, resources, or performance). The proper way to deal with uncertainty on a project is to conduct a thorough risk analysis, as described in the later sidebar "Working with risk."

For those risks you accept, because you either choose not to develop a response or can't further reduce their probability or impact, you can set aside contingency reserve to address the event when and if it occurs. No prescribed equation can determine the appropriate amount of reserve to set aside; every project is different. Here's a list of factors that generally indicate you should set aside more reserve:

- New or unproven technology
- Complex projects with numerous interfaces

- Projects that are unfamiliar to your organization
- Resources unfamiliar with the type of work to be done
- A constrained budget
- A business-critical project
- A high-profile project

To provide contingency reserve in the schedule, an unresourced task is commonly added at the end of every project phase or before a major deliverable is due. This strategy increases the likelihood that you'll meet the due date of a milestone or phase. Figure 12-1 shows how a task called Wall Contingency was added to reduce the probability that a slip on any task in the Walls work packages would delay the Walls Complete milestone. If everything goes as planned, then the contingency won't be used. However, if there is inclement weather, a delay in materials, or some tasks need to be reworked, there is some contingency time built in.

| Task Name | Duration | Predecessors |
|---|---|---|
| **Perimeter** | **70 days** | |
| **Walls** | **49 days** | |
| Dig trench | 2 wks | |
| Install footing | 1 wk | 3 |
| Install cinder blocks | 3 wks | 4 |
| Apply stucco | 1 wk | 5 |
| Paint | 1 wk | 6 |
| Apply finishes | 4 days | 7 |
| Wall contingency | 1 wk | 8 |
| Walls complete | 0 days | 9 |



**FIGURE 12-1:** Adding contingency in the schedule.

TIP

Set aside contingency reserve for both the schedule and the budget.

## Completing a task in less time

If you do your homework and add contingency reserve to the project (refer to the preceding section), you're making the plan realistic but also adding time to the project. When the project finish date just won't work for the powers that be, you have to try a few tactics to chop the timing down to size.

## WORKING WITH RISK

All projects have risk. There are too many unknown factors for you to be able to expect your project to go exactly as planned. To account for these unknown elements, take the time to follow these steps:

1. **Work with your team to create a list of all possible risks.**

   Analyze the WBS, schedule, budget, technical documentation, and any other documentation you can get your hands on.

2. **Calculate the probability that the risks you've identified will occur and determine the impact if they do occur.**

   For example, risks may affect the schedule, budget, scope, quality, stakeholder satisfaction, or another objective.

3. **Prioritize the risks by those that have the greatest impact and probability.**

4. **Develop a response plan for risks that are likely to have a significant impact if they occur.**

   Responses may include these strategies: Avoid the risk altogether, find a different way to perform a task, eliminate the risky part of the project, reduce the probability or impact of the event (or both), transfer the risk to someone else (such as a vendor) to handle, or accept the risk. If the risk is outside your authority, such as a risk that impacts multiple projects in a program or portfolio, then you can escalate the risk to someone with the authority to respond to it.

5. **Update the project schedule.**

   Include the work, resources, time, and funding that are necessary to implement the risk responses as appropriate.

## Checking dependencies

The timing of the plan is determined by the duration of each task and by its *dependencies* — the relationships you build between tasks. Ask yourself whether you've built all dependencies in the best way possible. Perhaps one task didn't start until another was finished but the second task could have started two days before the end of its predecessor. Building in this type of overlap (known as *fast tracking*) can save you time.

TIP

Use the Task Inspector pane, covered in Chapter 11, to scope out dependencies.

Here's an example: You created a finish-to-start relationship for the Do Research and Write Speech tasks, such that you couldn't start writing the speech until your research was finished. But is that true? Couldn't you do a first draft of the speech starting three-fourths of the way through the research? Especially when you have two different resources working on those tasks, initiating the second task before the first is complete can save you time.

Over the life of a project with hundreds of tasks, adding that kind of overlap to even a few dozen tasks can save you a month of time or more.

REMEMBER

You can take a refresher course in creating and changing dependencies in Chapter 4.

## Managing the availability of resources

Another factor that drives timing is the availability of resources. Sometimes in a dependency relationship, one task can't start before another ends, simply because the resources aren't available until the predecessor task is over. Look for these potential problems with resource-dependent timing:

- You delayed the start of a task because a resource wasn't available. But perhaps another resource can do the work. If so, switch resources and let the task start sooner.
- Project calculates the duration of auto-scheduled tasks (Fixed Work and Fixed Units with effort-driven scheduling) according to the number of resources available to do the work. If you add resources to these tasks, Project shortens their duration.
- If you assign a more highly skilled resource to certain tasks, you may be able to reduce the hours of work required to complete the task, because the skilled person can finish the work more quickly.
- Assigning more resources to tasks on the critical path can shorten those effort-driven tasks.
- You have money but no time or resources. Consider hiring an outside vendor to handle the work.

REMEMBER

Chapter 9 covers the mechanics of making and changing resource assignments.

## Cutting to the chase: Deleting tasks

When all else fails and you've reduced as much time as possible using the methods described in the preceding sections, it's time to cut some specific corners. First, consider skipping nonvital tasks. Then ask whether you can revisit the goals of the project with stakeholders and perhaps reduce its scope so that you can jettison tasks from the schedule. You can even negotiate for less-stringent quality or performance levels.

WARNING

*Never* eliminate all slack in the schedule. Slack is your friend; it provides scheduling flexibility around resources.

### Trading time for cost

Sometimes, the most critical objective in a project is the schedule. If that's the case on your project, you may be able to negotiate for a larger budget to reduce the overall duration. You then use the extra funds to pay for overtime, add resources, expedite shipping, or increase the hourly commitment of existing resources. Reducing the schedule by allocating money (in the form of resources) to the problem is called *crashing*. However, not all tasks can be crashed. Tasks that have truly fixed durations take as long as they take, regardless of the number of resources you throw at them.

To crash the schedule, start by analyzing tasks to see where you can deduct the most time from the schedule for the least cost. After you exploit this task, move to the next least-expensive task, and so on.

TIP

Focus only on the tasks that are on the critical path. Reducing the duration of tasks that aren't on the critical path does no good if you're trying to reduce the overall length of the project.

At some point, you'll reduce the duration of the critical path to the extent that you have a different critical path. Then you can go to work on the new critical path.

The more you fast-track, crash, and de-scope the project, the more likely it is to be late. That's right: At some point, you'll have compressed as much time as possible from the schedule and it will have multiple critical paths and near-critical paths. If any task on any of those paths slips, the project will need to use some contingency reserve. And if you aren't careful, you'll use it all up and make the project late.

## Getting What You Want for Less

After you assign all resources to tasks and set all fixed costs, sticker shock may set in. Project tallies all those costs and shows you the project's budget. If those numbers simply won't work, follow these tips for trimming the bottom line:

- **Use less expensive resources.** A high-priced engineer may be assigned to a task that can be performed by a junior engineer, for example. Or you may have assigned a high-priced manager to supervise a task that can be handled by a lower-salaried line supervisor.
- **Lower fixed costs.** If the budget allows for the travel costs associated with four plant visits, calculate whether you can manage with only three, and try to book flights ahead of time to get cheaper airfares. Look for a vendor who can sell you that piece of equipment for less than the $4,000 you allocated — or make the old equipment last for one more project.
- **Reduce overtime.** If resources who earn overtime are overallocated, reduce their hours or use resources who are on straight salary for those 14-hour days.
- **Complete the task in less time.** The resource cost is a factor of task duration, hourly wage, or number of units. If you change tasks so that fewer work hours are needed in order to complete them, tasks cost less. However, don't be unrealistic about the time it *truly* takes to complete the work.

# The Resource Recourse

Before you finalize the plan, consider resource workload. As you delved into assigning resources to tasks in the project, you probably created situations where resources must work around the clock for days on end. The first step in finalizing your resources is to spot those overallocations — then you have to help those poor folks out!

## Checking resource availability

To resolve problems with resource assignments, you have to first figure out where the problems lie by taking a look at a few views that focus on resource assignment.

Resource Usage view, shown in Figure 12-2, and Resource Graph view, shown in Figure 12-3, are useful in helping you spot overbooked resources. In Resource Usage view, the overbooked resource is shown in bold red, and the overallocation icon (a little red person in the Indicator column) appears. The total hours that the resource is working each day on the combined tasks are summarized on the line that lists the resource's name. In this case, Vijay is overallocated for four days.

| Vijay Chandra | 220 hrs | Work | 12h | 12h | 12h | 12h | 8h |
|---|---|---|---|---|---|---|---|
| *Purchase Trucks* | *60 hrs* | Work | | | | | |
| *Purchase Carts* | *60 hrs* | Work | | | | | |
| *Purchase Radios* | *60 hrs* | Work | 4h | 4h | 4h | 4h | |
| *Purchase Phones* | *20 hrs* | Work | 4h | 4h | 4h | 4h | 4h |
| *Purchase Tablets* | *20 hrs* | Work | 4h | 4h | 4h | 4h | 4h |

**FIGURE 12-2:** Resource Usage view.

*© John Wiley & Sons, Inc.*

In Resource Graph view, the graph bars are shown in red for any allocation over the units allowed for the resource. In this view, work is summarized in the Peak Units row at the bottom, and all work that exceeds the maximum allowable amount is shown in red in the bar graphic.

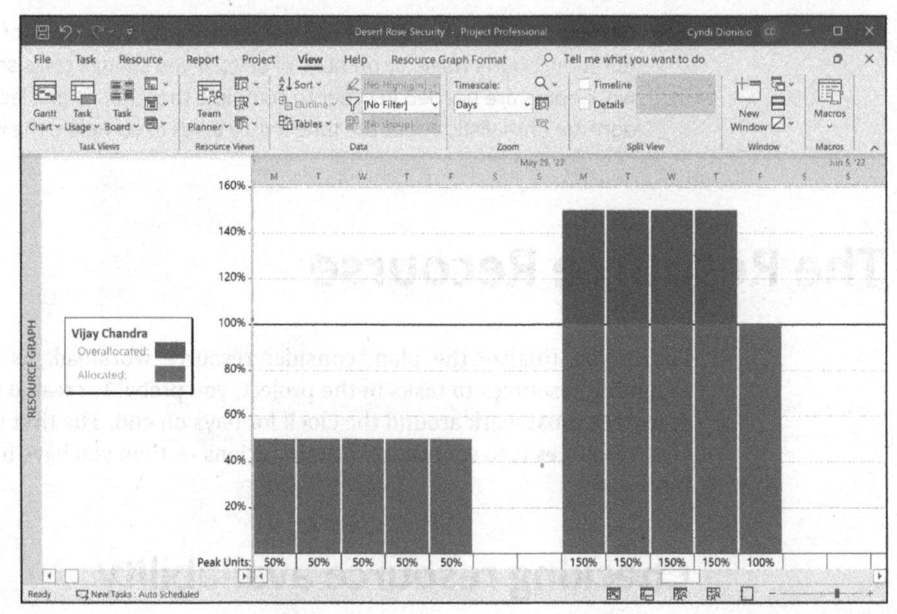


**FIGURE 12-3:** Resource Graph view.

*© John Wiley & Sons, Inc.*

Keep in mind that resources are flagged as overallocated in these Resource views based on their assignment percentages and calendars. A resource based on a standard calendar showing 8-hour days, assigned at 100 percent to a task, will work 8 hours a day on it. If you assign the same resource at 50 percent to another task that happens at the same time, the resource will put in 12-hour days (8 plus 4) and be marked as overallocated.

## Deleting or modifying a resource assignment

Suppose that you discover that a resource (I'll call her Janet) is working 16 hours on Tuesday and 14 hours on Friday. You have several options to rectify the workload:

- **Remove Janet from a few tasks to free up some time.** This option is good if she's working on noncritical tasks, especially if other resources are working on those same tasks.
- **Change Janet's Resource calendar to allow for a longer workday — for example, 12 hours.** A 100 percent assignment will therefore have Janet working 12 hours on any single task in a day. If you stretch a resource's workday, reduce the resource's assignments also. For example, if someone frequently spends 16 hours on two tasks in one day (based on an 8-hour calendar) and two 100 percent assignments, change to a 12-hour calendar and 50 percent assignments (6 hours on each of the two tasks, totaling 12 hours a day). However, if the resource typically works an 8-hour day and 12- or 16-hour days *are the exception,* don't change the resource's base calendar; it will affect *all* assignments for that resource. And if an overtime pay rate applies, remember to authorize (add) it.

WARNING

The first two options in this list may lengthen the tasks that the resource is assigned to, regardless of whether you remove the resource or reduce the resource's assignment.

- **Change Janet's availability by increasing her assignment units to more than 100% in the Resource Information dialog box.** For example, if you enter **150%** as her available units, you're authorizing her to work 12 hours a day, and Project then doesn't consider her overallocated until her working time exceeds 12 hours.
- **Use Team Planner view to shuffle some of Janet's tasks to other resources.** When you're making the switch, Team Planner view provides the best visibility into how changes affect various resources.
- **Ignore the problem.** I don't mean this statement facetiously; sometimes, a resource can reasonably work 14 hours *for a day or two* during the life of a project, and you don't need to change the resource's usual working allocation to eliminate the exclamation point in the indicator column. (However, let Janet order pizza on the company's tab on those long workdays — and don't let these long days become the norm.) A resource who is scheduled for even a few minutes more than the calendar allows is flagged as overallocated. As the project manager, it isn't worth the effort for you to attempt to resolve tiny overallocations that don't affect the project schedule.

## Beating overallocations with quick-and-dirty rescheduling

Gantt Chart view also clues you in to overbooked resources. When a task has an overallocated resource assigned, the Overallocation icon appears in the Indicator column. To fix the overallocation, you can right-click the icon and choose Move Task to Resource's Next Available Time or you can select Fix in Task Inspector (refer to Figure 11-11 in the preceding chapter). Moving the task to the resource's next available time reschedules all or part of the task to a period when the resource is available to fix the overallocation, if possible. You can always undo the change if you disagree with its impact on the overall project schedule. This method can't fix all overallocations, so if the icon remains after you choose Reschedule to Available Date, look at other methods to resolve the problem. Right-clicking the icon again and choosing Fix in Task Inspector from the menu enables the Task Inspector to give you ideas for fixing the overallocation.

## Finding help

When one person is overworked, you have to look for help. You can free resources in several ways.

For starters, you can assign someone to help on a task and thus reduce the overbooked resource's workload. Reduce the resource's work assignment on one or more tasks, perhaps by reducing 100 percent assignments to 50 percent. You do this in the Resources Tab in the Task Information dialog box or by selecting the task and clicking the Assign Resources button on the Resource tab of the Ribbon to open the Assign Resources dialog box.

TIP

Quickly check a work graph for any resource by clicking the Graph button in the Assign Resources dialog box.

You'll also find that by adding resources to certain tasks, you shorten the task duration. You may, therefore, free the resource in time to eliminate a conflict with a later task in the project.

Experiment with changing the work contour of the resource. By default, Project has a resource work on a task at the same level, or *flat*, for the life of the task. You can modify the work contour, for example, so that a resource invests the highest level of effort at the start of a task (front loaded), which reduces the resource's workload later when a conflict with another assignment might occur. See Chapter 9 for information about how to apply a work contour to a resource assignment.

## Leveling resources

Project calculates *resource leveling* to try to resolve resource overallocations in the project. The feature works in two ways: by delaying a task until the overbooked resource is freed up or by splitting tasks. Splitting a task involves (essentially) stopping it at some point, thereby freeing the resource, and then resuming it later, when the resource is available.

You can make these types of changes yourself or let Project make the calculation. Project first delays tasks that involve overallocated resources to use up any available slack. When no more slack is available on these tasks, Project makes changes based on any priorities you've entered for tasks, dependency relationships that are affected, and task constraints (such as a Finish No Later Than constraint).

Don't worry: You can turn on leveling to see which changes Project would make and then clear the leveling to reverse those actions if you don't like the results.

To level the resources in the project, follow these steps:

1. **From the Resource tab, go to the Level group and select Leveling Options.**

   The Resource Leveling dialog box appears, as shown in Figure 12-4.

2. **Choose whether to allow Project to level automatically or manually:**
   - *Automatic* tells Project to level every time you change the plan.
   - *Manual* requires you to click the Level All button in the Resource Leveling dialog box or use the Level All button on the Resource tab.

3. **If you choose to level automatically, select the Clear Leveling Values before Leveling check box if you want previous leveling actions to be reversed before you level the next time.**

4. **Set the leveling range to one of these options:**
   - *Level Entire Project*
   - *Level <a date range>*

   If you choose the latter option, fill in a date range by making choices in the From and To boxes.

5. **From the Leveling Order drop-down list, click the down arrow and make a choice:**
   - *Standard* considers slack, dependencies, priorities, and constraints.
   - *ID Only* delays or splits the task with the highest ID number — in other words, the last task in the project.

- *Priority, Standard* uses task priority as the first criterion in making choices to delay or split tasks (rather than using up slack). You can set task priority in the Task Information dialog box. All tasks are set to a default value of 500 (out of 1,000).

6. **Select any of the five check boxes at the bottom to control how Project levels:**

   - *Level Only within Available Slack:* No critical tasks are delayed, and the current finish date for the project is retained.
   - *Leveling Can Adjust Individual Assignments on a Task:* Project removes or changes assignments.
   - *Leveling Can Create Splits in Remaining Work:* Certain tasks are placed on hold until resources are freed up for work.
   - *Level Resources with the Proposed Booking Type: Booking type* (proposed or committed) relates to the firmness of your commitment to using that particular resource. Allowing resource leveling to consider a resource's booking type means that committed resource assignments are considered more sacred than proposed assignments when Project makes changes.
   - *Level Manually Scheduled Tasks:* Project moves the task even if you've scheduled it manually.

7. **Click the Level All button to have Project perform the leveling operation.**

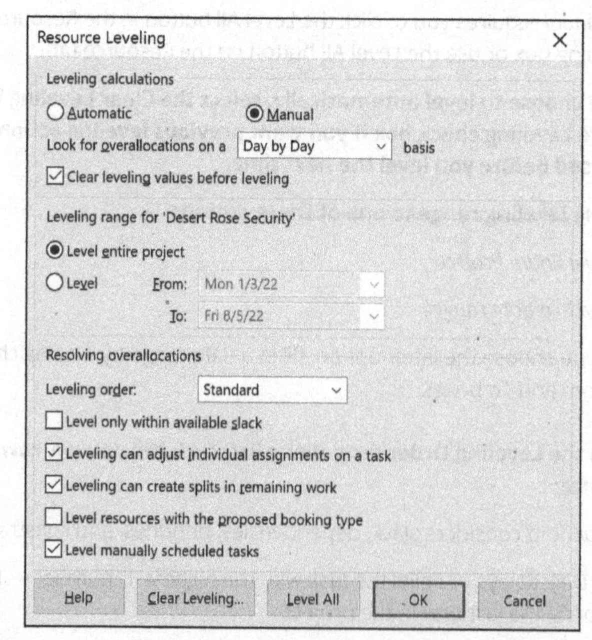


**FIGURE 12-4:** The Resource Leveling dialog box.

TIP

**To reverse leveling, choose Resource Tab⇨Clear Leveling from the Ribbon.**

**When you have only a few instances of overallocated resources, go to the Resource tab and choose Level Group⇨Level Resource. Select the resource from the Level Resources dialog box (shown in Figure 12-5) and then click Level Now.**

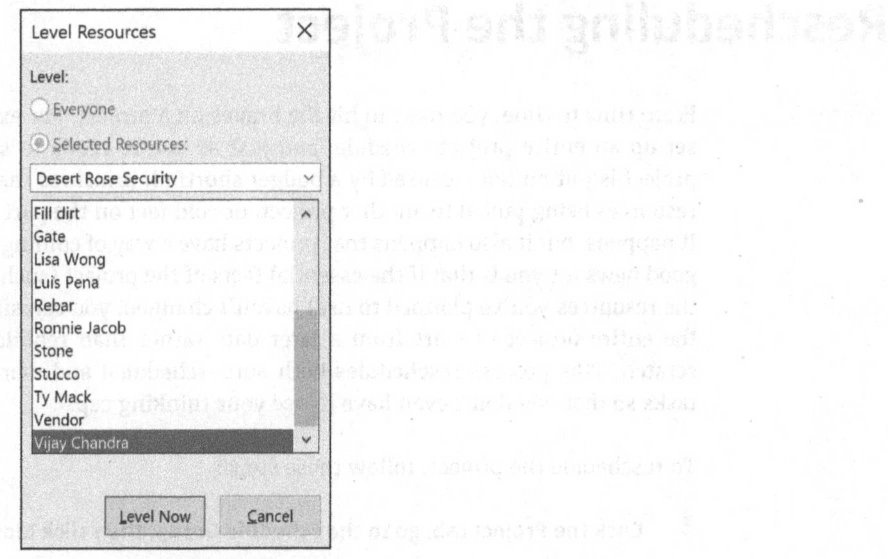


**FIGURE 12-5:** The Level Resources dialog box.

*© John Wiley & Sons, Inc.*

## TO LEVEL OR NOT TO LEVEL?

As both a process and a program feature, resource leveling has pros and cons. It can make changes that you may not want it to make — for example, removing a resource from a task where you absolutely need that person's unique skills. Leveling frequently delays the project's finish date, which may not be acceptable to you (or your boss).

The safest setting for resource leveling (the one that makes the least drastic changes to the timing) is to level only within slack. This setting may delay certain tasks, but it doesn't delay the project completion date.

If you can't live with the changes resulting from resource leveling, the capability to turn it on and off is vital. After you turn on this feature, look at its changes to resolve resource problems, and then turn it off and manually institute the portions of the solution that work for you.

REMEMBER

To be successful, take the time to find the best combination of all these methods. Solving problems is often a trial-and-error process. Although you may initially look for a single quick fix, the best solution usually results from making lots of small changes.

# Rescheduling the Project

From time to time, you have to hit the brakes on a project. For example, you may set up an entire project schedule, and just as you're ready to start rolling, the project is put on hold, caused by a budget shortfall, priorities that have changed, resources being pulled to another project, or cold feet on the part of stakeholders. It happens, but it also happens that projects have a way of coming back to life. The good news for you is that if the essential facts of the project (such as its scope and the resources you've planned to use) haven't changed, you can simply reschedule the entire project to start from a later date rather than rebuild the plan from scratch. This process reschedules both auto-scheduled and manually scheduled tasks so that you don't even have to use your thinking cap.

To reschedule the project, follow these steps:

1. **Click the Project tab, go to the Schedule Group, then click Move Project.**

   The Move Project dialog box appears, as shown in Figure 12-6. You can use it to reschedule an entire project that was put on hold.

2. **Enter the new project start date in the New Project Start Date field, or choose a date from the drop-down calendar.**
3. **Select the Move Deadlines check box if you want Project to move any deadlines that are assigned to tasks.**
4. **Click the OK button.**

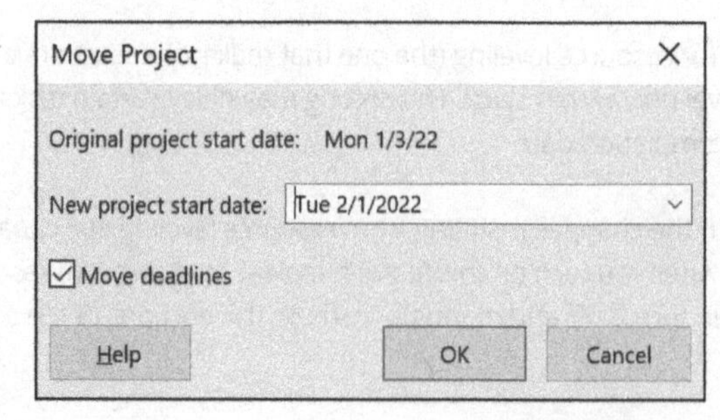


**FIGURE 12-6:** The Move Project dialog box.

© John Wiley & Sons, Inc.

IN THIS CHAPTER

» Formatting the Gantt chart

» Formatting Network Diagram task boxes

» Dressing up task boxes

» Modifying the layout of a view

» Changing the gridlines in a view

» Adding graphics to the project

» Creating a custom field

# Chapter 13 Making the Project Look Good

You may know the saying "Clothes make the man." Well, in the same spirit, sometimes the look of the schedule makes the project. Having a project schedule that looks good serves two purposes: It impresses people with your professionalism and it makes it easy for others to discern the different boxes, bars, and lines.

Project uses default formatting that's helpful in most cases. However, if you have certain company standards for reporting — say, representing baseline data in green and actual data in blue, or more frequent gridlines to help the nearsighted CEO read Project reports more easily — Project has you covered.

Whatever you need, Project provides tremendous flexibility in formatting various elements in the plan.

## Looking Good!

Microsoft has decided to capitulate to the artist in all of us by allowing users to modify shapes, colors, patterns, and other graphical elements in the Project schedule. You gain great flexibility in determining how the schedule looks.

When you print a Project view (as covered in Chapter 18), you can print a legend on every page. The legend helps readers comprehend the meaning of the various colors in the plan (or the shading, if it's printed in black and white) along with shapes that you set for elements.

All the views and formatting choices that Project offers you aren't confined to the screen; you can print the schedule or reports. Whatever information is displayed onscreen when you print a view is also printed. So knowing how to make all kinds of changes onscreen allows you to present information to team members, managers, vendors, and clients on hard copy too.

Printing in color is useful because you can provide the full visual impact and nuance of the various colors used for graphical elements, such as taskbars and indicators. If you print in black and white, you may find that certain colors that look good onscreen aren't distinct when you print them. When you modify the formatting, the project can look good in color *and* in black and white, and onscreen and in print.

## Formatting the Gantt Chart

Project allows you to customize the Gantt chart in several ways. You can choose different taskbars, add text inside taskbars, and supply different bar ends. You can highlight the critical path so that it stands out, and you can change the font styles in the sheet section of the schedule.

### Formatting taskbars

*Taskbars* are the horizontal boxes that represent the timing of tasks in the Chart pane of Gantt Chart view. You can format a bar individually, change the formatting settings on different types of taskbars, or apply a new style that controls bar formatting on the entire chart.

You can change several characteristics about taskbars:

- **The shape that appears at the start and end of the bar:** The ends can be formatted with different shapes, such as arrows and triangles.
- **The middle of the bar:** You can change the shape's type, pattern, and color.
- **The text that you can set to appear in five locations around the bar (to the left or right of the bar or above, below, or inside it):** You can include text in any or all of these locations; just don't add so many text items that they become impossible to read. As a rule, use only enough text to help readers of the plan identify information, especially on printouts of large schedules where a task may appear far to the right of the Task Name column that identifies it by name in the sheet area.

When you track progress on a task, a progress bar is superimposed on the taskbar. You can format the shape, pattern, and color of the progress bar. The goal is to contrast the progress bar with the baseline taskbar so that you can see both clearly.

WARNING

By formatting taskbars, you can help readers of the plan identify various elements, such as progress or milestones. If you make changes to individual taskbars, people who are accustomed to the standard formatting in Project may have trouble reading the plan. In other words, don't get carried away.

To create formatting settings for various types of taskbars, follow these steps:

1. **From the Gantt Chart Format contextual tab, select Format in the Bar Styles group and then choose Bar Styles.**

   The Bar Styles dialog box appears, as shown in Figure 13-1.

2. **In the Name column, select the type of task that you want to modify (Task, Split, Summary, or Milestone, for example).**

   To modify the styles used for all summary tasks, for example, select Summary. The choices on the Bars tab in the bottom half of the dialog box change based on the selected task type.

3. **Click in the Show For Tasks column for the task type you want to modify.**

   A drop-down list appears, as shown in Figure 13-2.

4. **Select the criteria for the task, such as Critical or Finished, from the drop-down list.**

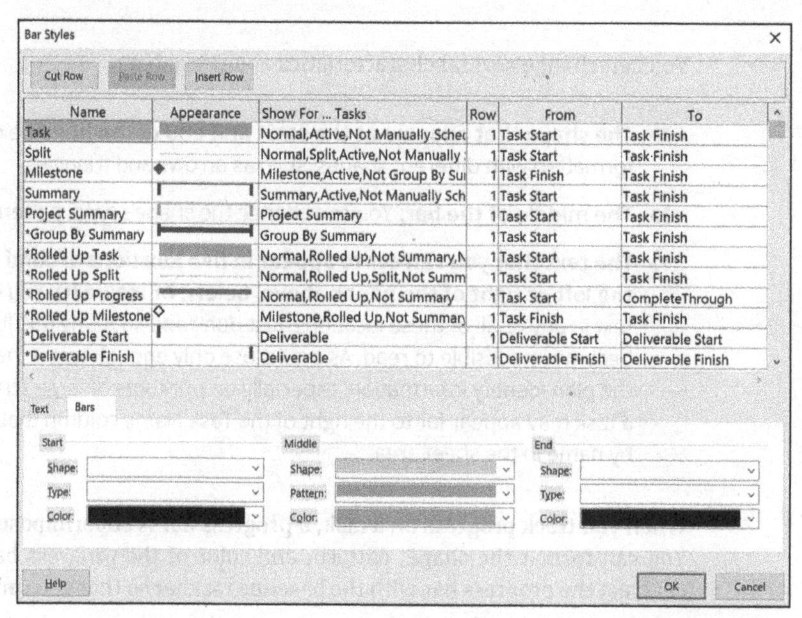


**FIGURE 13-1:** The Bar Styles dialog box.



5. **Click the Bars tab at the bottom of the dialog box to display your options, and then make selections as needed from the following drop-down lists:**

   - *Shape:* Modify the shape of either end or the middle of the taskbar.

     Shapes on either end may be any one of 24 shapes, though you may commonly see an arrow, a diamond, or a circle. The shape in the middle consists of a bar of a certain width.

   - *Type:* Modify the type of formatting for the shape on either end of the taskbar.

     This setting controls how the shape is outlined: framed with a solid line, surrounded by a dashed line, or filled in with a solid color.

   - *Pattern:* Select another pattern for the middle of the bar.

   - *Color:* Modify the color used on either end or in the middle of the taskbar.

     Project also allows you to choose among various theme colors (as in the other Office applications).

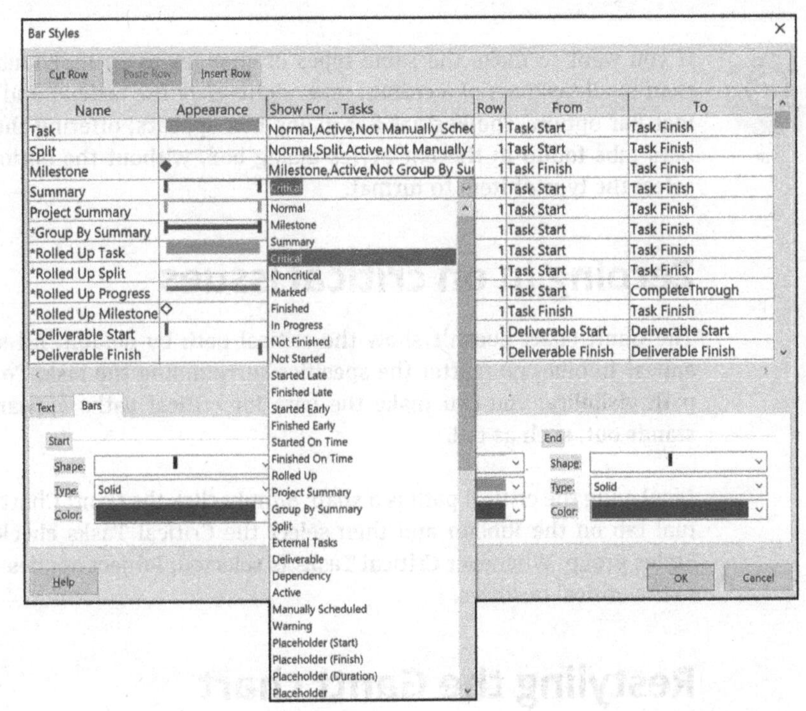


**FIGURE 13-2:** The Show For Tasks drop-down list.

© John Wiley & Sons, Inc.

6. **Click the Text tab and then:**

   *(a) Click any text location.*

   A drop-down button appears at the end of that line.

   *(b) Click the drop-down button to display an alphabetical list of possible data fields to include and then click a field name to select it.*

   *(c) Repeat Steps 6a and 6b to choose additional text locations.*

   Figure 13-3 shows how resource names are selected for the right side of the bar and how the WBS is selected inside the bar.

7. **Click the OK button to accept all new taskbar settings.**

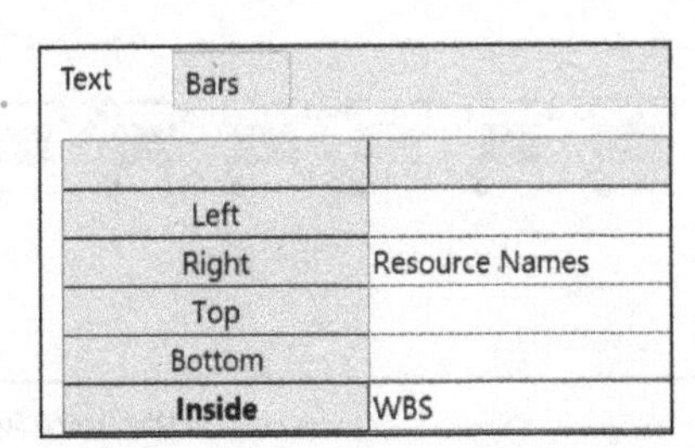


**FIGURE 13-3:** Add text to taskbars.

© John Wiley & Sons, Inc.

TIP

If you want to make the same types of changes to an individual taskbar rather than to all taskbars of a certain type, right-click the taskbar and choose the Format Bar option. The Format Bar dialog box appears, offering the same Text and Bars tabs found in the Bar Styles dialog box, without the options at the top to select the type of item to format.

## Zeroing in on critical issues

The Gantt chart doesn't show the critical path by default. All subtask taskbars appear in blue, no matter the specifics surrounding the task. To give the critical path visibility, you can make the bars for critical paths appear in a color that stands out, such as red.

Displaying the critical path is a snap: Simply click the Gantt Chart Format contextual tab on the Ribbon and then select the Critical Tasks check box in the Bar Styles group. Whenever Critical Tasks is selected, Project applies the special color to the critical taskbars.

## Restyling the Gantt chart

If those boring red and blue Gantt bars are cramping your style, not to worry: You can try a new look, including a different color scheme, by applying a new Gantt Chart style. Working in Gantt Chart view, click the Gantt Chart Format contextual tab. Then click the More drop-down button (a horizontal line over a downward triangle) in the Gantt Chart Style group to display the available styles, as shown in Figure 13-4. Select the style you want, and Project reformats the whole project. It doesn't get much simpler!

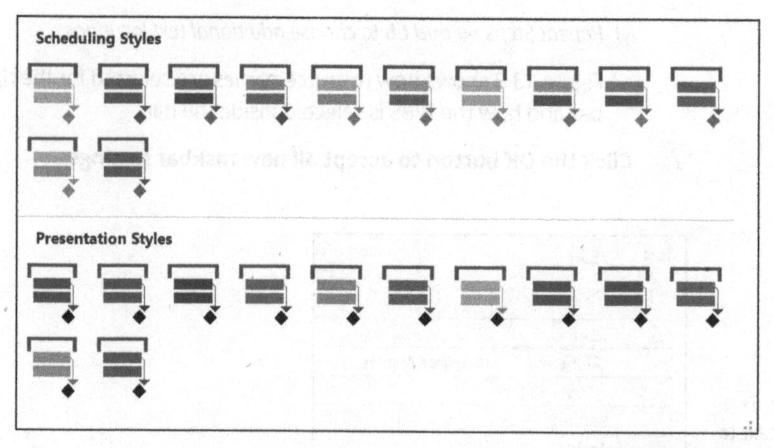


**FIGURE 13-4:** Gantt chart styles.

If you want to format the sheet portion of Gantt Chart view, too, start by clicking the Gantt Chart Format contextual tab. Click the Text Styles button in the Format group to open the Text Styles dialog box for changing the text font, size, and other options.

Figure 13-5 shows a customization of critical tasks to have a bold Calibri 12-point font with a purple color. You can choose any font, style, size, color, pattern, or background for the font style. You aren't limited to changing only critical tasks; you have 18 types of tasks you can change, including milestones, summary tasks, and bar text styles.

You can also use the Wrap Text button in the Columns group to control whether text automatically wraps in a particular cell or column.

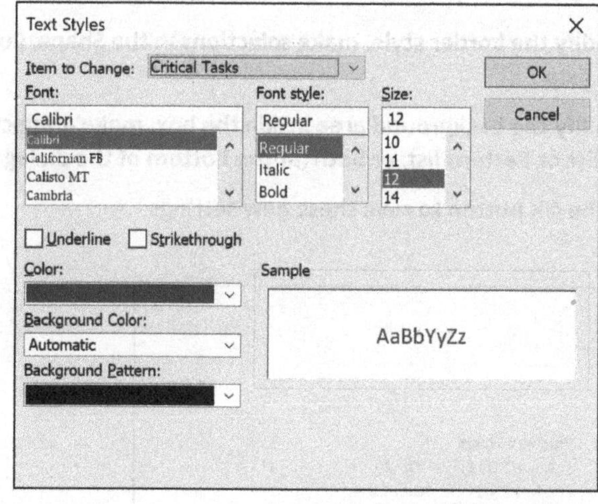


**FIGURE 13-5:** Customizing critical tasks in the Text Styles dialog box.

*© John Wiley & Sons, Inc.*

# Formatting Network Diagram Boxes

Network Diagram task boxes use different shapes to help you spot different types of tasks:

- **Summary tasks** use a slightly slanted box shape and include a plus sign (+) or minus symbol (–), depending on whether the summary task's subtasks are hidden or displayed. Click the symbol to hide or display subtasks.
- **Subtasks** show up in simple rectangular boxes.
- **Milestones** are shown in boxes with points on both ends.

You can change the formatting of each task box individually or by type. To change the formatting of task boxes displayed in Network Diagram view, follow these steps:

1. **Display Network Diagram view.**

   From the Task tab, go to the View group and select the down indicator. Choose Network Diagram.

2. **Right-click the task box that you want to change and then choose Format Box.**

   Alternatively, right-click anywhere outside the task boxes and then choose Box Styles to change the formatting of all boxes of a certain style.

   The Format Box dialog box (or the Box Styles dialog box) appears, as shown in Figure 13-6.

3. **To modify the border style, make selections in the Shape, Color, and Width lists.**
4. **To modify the background area inside the box, make a selection in the Color list or Pattern list, or both (at the bottom of the dialog box).**
5. **Click the OK button to save these new settings.**

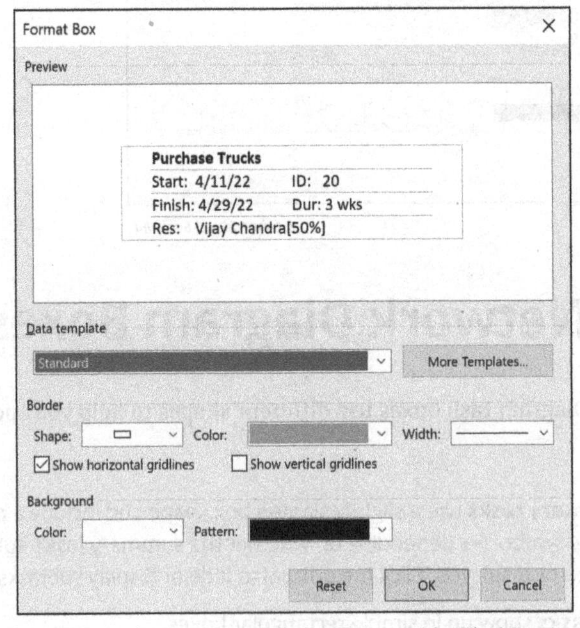


**FIGURE 13-6:** The Format Box dialog box.

WARNING

When you modify the formatting of individual boxes in Network Diagram view, the standard settings of slanted summary tasks no longer serve as visual guides to *all* tasks. If you make changes and decide to restore a task box to its default setting, click the Reset button in the Format Box dialog box.

# Adjusting the Layout

In addition to displaying particular columns and formatting taskbars, you can make certain changes to the layout of the view. These options vary a great deal, depending on the view. The layout of Calendar view and Network Diagram view is quite different from the layout choices offered in Gantt Chart view, for example.

To display the Layout dialog box for a view, click the Format contextual tab and, in the Format Group, select Layout. You can also right-click the area (for example, the chart area in Gantt Chart view) or anywhere in Calendar view or Network Diagram view and then choose Layout from the menu that appears. Figures 13-7 and 13-8 show the various layout choices available for the Gantt Chart and Network Diagram views, respectively.

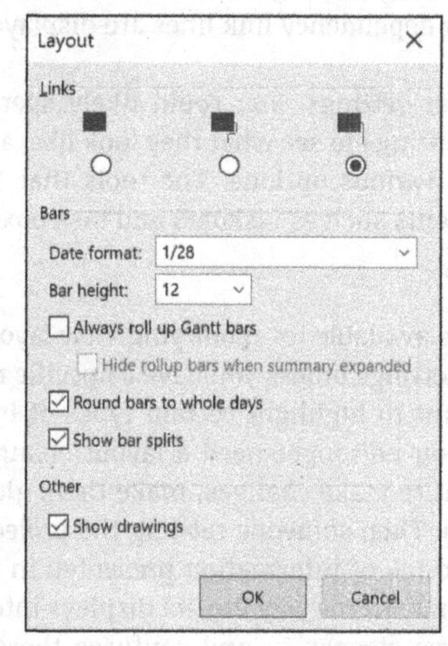


© John Wiley & Sons, Inc.

**FIGURE 13-7:** The Layout dialog box in Gantt Chart view.

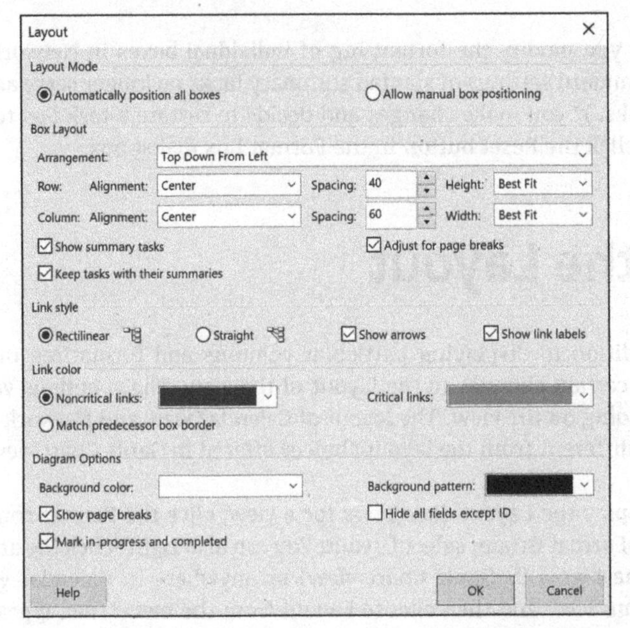


**FIGURE 13-8:** The Layout dialog box in Network Diagram view.



The settings in these Layout dialog boxes generally reflect how the elements on the page are arranged and how dependency link lines are displayed.

Table 13-1 describes the layout settings. You could likely spend a week or so experimenting with all these settings to see what they look like, and I could spend a few days writing about the various options. The tools that Project offers to modify the formatting of elements such as taskbars and task boxes provide helpful flexibility.

Now that you know the options available for modifying view layouts, I have some advice: Stick with the default settings unless you have a specific reason to make a change (such as when you want to highlight certain types of information for a project presentation). When you no longer need a layout change, return to the default settings. Or if you want to make changes, make them globally across the organization and stick to them. Then someone reading the project plan can more easily interpret the different kinds of information presented in views. Generally speaking, if you tinker too often with the way Project displays information, it only makes your own learning curve steeper — and confuses those who know the default settings in Project.

**TABLE 13-1 Layout Options**

| Layout View | Option | How It's Used |
|---|---|---|
| **Network Diagram** | Layout Mode | Automatic or manual positioning is allowed. |
| | Box Layout | These settings arrange and align boxes; adjust alignment, spacing, and height; and modify the display of summary tasks. |
| | Link Style | The style for dependency link lines is modified. |
| | Diagram Options | The background color and pattern of boxes as well as page breaks and task progress are controlled. |
| **Gantt Chart** | Links | The style of lines indicating dependency links are set. |
| | Date Format | The format of the date used to label taskbars is modified. |
| | Bar Height | The height of taskbars is set in points. |
| | Always Roll Up Gantt Bars | When this option is selected, taskbar details roll up to the highest-level summary task. |
| | Round Bars to Whole Days | When you use portions of days, bars are allowed to represent the nearest whole-day increment. |
| | Show Bar Splits | A task that includes a period of inactivity can be shown as split into different segments over time. |
| | Show Drawings | Drawings are displayed onscreen and on the printout. |

# Modifying Gridlines

Just as phone numbers are divided into several shorter sets of numbers to help you remember them, visual elements are often broken up graphically to help you understand information in chunks. Tables use lines, calendars use boxes, football fields use yard lines, and so on.

Several views in Project include gridlines to indicate certain elements, such as a break between weeks or the status date (that is, the date to which progress has been tracked on a project). These lines help someone reading the plan to discern intervals of time or breaks in information; for example, gridlines can be used to indicate major and minor column breaks. You can modify these gridlines in several ways, including changing the color and style of the lines and the interval at which they appear.

To modify gridlines, use the Gridlines dialog box and follow these steps:

1. **Right-click any area of a view that contains a grid (for example, the chart area of Gantt Chart view or Calendar view) and then choose Gridlines.**

   The Gridlines dialog box appears, as shown in Figure 13-9. You can also display the Gantt Chart Format contextual tab, select Gridlines in the Format group, and then click Gridlines to open the dialog box.

2. **In the Line to Change list, select the gridline you want to modify.**
3. **In the Normal grouping, use the Type drop-down list and the Color drop-down list to select a line style and color.**
4. **If you want to use a contrasting color at various intervals in the grid to make it easier to read, follow these steps:**

   (a) *Select an interval at which to include a contrasting line.*

   This setting is typically used with a different style or color from the Normal line setting to mark minor intervals for a grid. Note that not every type of gridline can use contrasting intervals.

   (b) *Choose the type and color of that line from the applicable drop-down lists.*

5. **Click the OK button to save these settings.**

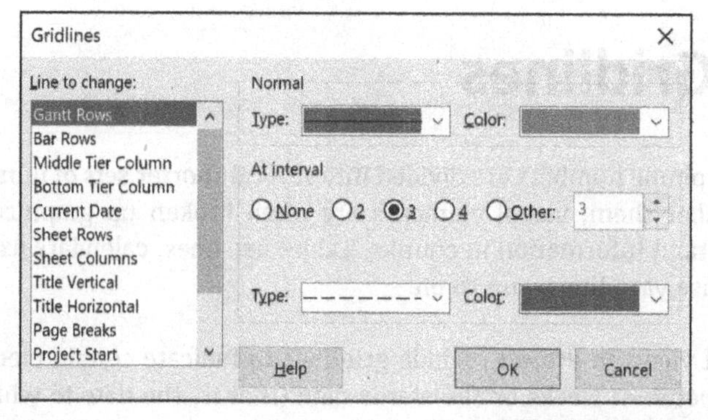


**FIGURE 13-9:** The Gridlines dialog box.

© John Wiley & Sons, Inc.

REMEMBER

**You make choices for modifying gridlines one by one, and you have no Reset button to revert to the original settings. Gridlines modified in one view don't affect gridlines in any other view.**

## Recognizing When a Picture Can Say It All

Words, numbers, taskbars, and task boxes work well in the plan, but you may want to add elements of your own, such as drawing a circle around a task to draw attention to it or adding a simple drawing to show a process or working relationship.

You can use a drawing tool on the Gantt Chart Format contextual tab of the Ribbon to draw images in the chart area of Gantt Chart view. Follow these steps to add a drawing:

1. **Display Gantt Chart view.**
2. **Click the Gantt Chart Format tab and then, in the Drawings group, click Drawing.**

   The menu of available shapes and commands appears, as shown in Figure 13-10.
3. **Click the drawing tool that represents the type of object you want to draw, such as an oval or a rectangle.**
4. **At the location on the chart where you want to draw the object, drag the mouse until the item is drawn approximately to the scale you want.**
5. **Release the mouse button.**

TIP

When you draw an object over an element such as a taskbar, the drawn object is solid white and covers up whatever is beneath it. You have a couple of choices if you want the item underneath to show through:

- Right-click the object, choose Properties, and click None in the Fill section of the Format Drawing dialog box to remove the fill.
- Right-click the object, choose Arrange, and choose Send to Back to place the object behind the chart elements.

At this point, you have these options:

- **Add text:** If you draw a text box, you can click in it and then type whatever text you want.
- **Resize objects:** Resize any object you've drawn by selecting it and then dragging any of the resize handles (the little black boxes) around its edges outward to enlarge it or inward to shrink it.
- **Move objects:** Move the mouse over the object until a four-way arrow appears below the pointer. Then drag the object elsewhere on the chart.

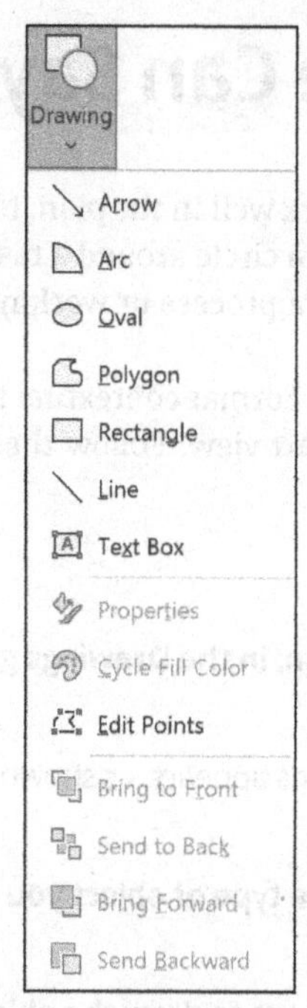


**FIGURE 13-10:** Drawing tools with basic shapes.

© John Wiley & Sons, Inc.

## Creating a Custom Text Field

The tables that Project offers for sheet views often restrict you to particular types of entries. For example, you have to enter a numerical value in any cost-related field. Project offers a number of placeholder fields that you can customize for your own purposes and add to any task or resource table. You can customize the text, cost, number, flag, and other types of fields. A neat shortcut is to create a custom field with a lookup table.

You can use a *lookup table* to create a drop-down list of values to select to create a custom field. That is, when you click a cell in the field, you can then click the drop-down list arrow that appears and select predefined entries. Suppose that you want to add a field that indicates whether an invoice has been submitted, paid, or posted. Creating a custom field with the lookup table offering the possible states of payment can make data entry faster and also help prevent data-entry mistakes that can occur if the field has no restrictions. By default, only the entries in the lookup table may be entered for the field.

To create and use a custom text field with a lookup table, follow these steps:

1. **Display the task- or resource-oriented view and then display the table that you want to customize.**

   You display tables by choosing Tables from the View tab and selecting a table from the list that appears.

2. **Right-click any field column heading and choose Custom Fields from the menu that appears.**

   The Custom Fields dialog box appears.

3. **Select the type of field you want from the list of options, as shown in Figure 13-11.**

4. **Click Rename, type a name for the field in the Rename Field dialog box, and click the OK button.**

5. **Back in the Custom Fields dialog box, click the Lookup button.**

   The Edit Lookup Table dialog box appears.

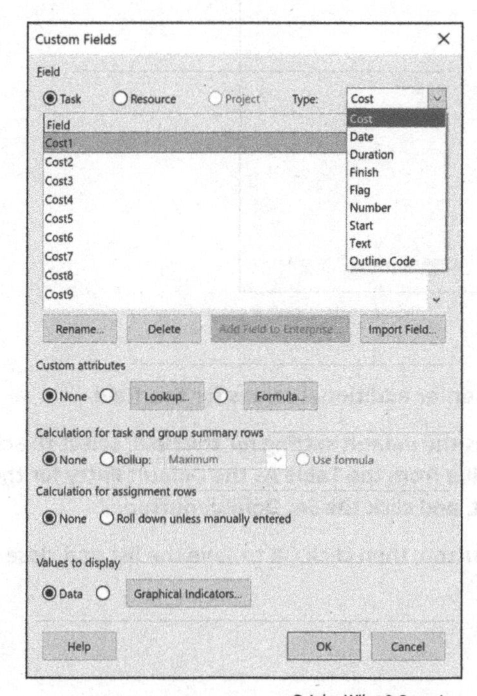


**FIGURE 13-11:** Custom field types.

© John Wiley & Sons, Inc.

6. **Enter a possible value (such as a date for a date field or a dollar amount for a cost field) in the Value column.**

   You can enter a description for each value in the Description column if necessary. Figure 13-12 shows values and descriptions for an invoice status lookup table.

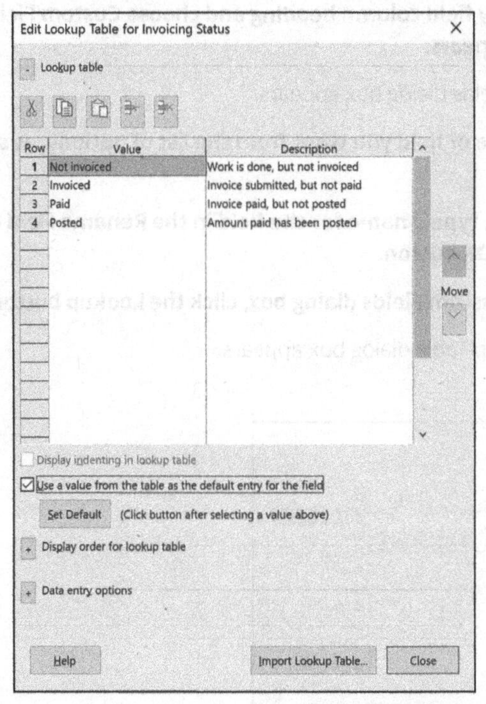


**FIGURE 13-12:** Setting values for the field.

© John Wiley & Sons, Inc.

7. **Repeat Step 6 to enter additional values for this field.**
8. **To set an entry as the default setting for the field, select the check box labeled Use a Value from the Table As the Default Entry for the Field, click a value in the list, and click the Set Default button.**
9. **Click the Close button; then click OK to save the list and close all dialog boxes.**

10. **Add the field to the location you want in the table.**

    Click the column to the right of where you want to insert the new column. Choose Format ➪ Insert Column and then select the column from the drop-down menu that appears.

    Figure 13-13 shows a custom text field with a lookup table added.

| Task Name | Remaining | Invoicing Status | June 2022 5/22 5/29 6/5 6/12 6/19 6/ |
|---|---|---|---|
| **Perimeter** | **$39,725.00** | | |
| **Walls** | **$16,600.00** | | |
| Dig trench | $3,000.00 | | |
| Install footing | $2,000.00 | Not invoiced Work is done, but not invoiced | |
| Install cinder blocks | $6,000.00 | Invoiced Invoice submitted, but not paid | |
| Apply stucco | $2,000.00 | Paid Invoice paid, but not posted | |
| Paint | $2,000.00 | Posted Amount paid has been posted | |



**FIGURE 13-13:** The Value list from the Invoicing Status field.

IN THIS CHAPTER

» Saving plan information with a baseline

» Using multiple baselines

» Setting and resetting baselines

» Saving interim plans

» Clearing and resetting interim plans

# Chapter 14
# It All Begins with a Baseline

Before you start tracking your progress, you need a benchmark to compare your status with so you know if you're ahead or behind schedule. In project management, we call this benchmark a *baseline*. A schedule baseline is the agreed-upon version of the schedule, against which you'll measure progress. It includes all information in the project, such as task timing, resource assignments, and costs.

An *interim plan* in Project is essentially a timing checklist. It includes the start and finish dates of tasks, and the estimated start and finish dates for tasks not yet started.

This chapter shows you when, why, and how to save a baseline and an interim plan for the project.

# All about Baselines

After you complete all the planning necessary to develop a robust schedule (and other documents) for a project, and after you've made the schedule presentable, it's time to baseline it. The *baseline* represents the project at the moment you consider the plan final and you're ready to commit to it before you begin any activity. The baseline, which is saved in the original Project file, exists alongside the progress you record on tasks in the project.

Key data that a baseline captures includes the baseline start, finish, duration, work, and cost information for each task. After you save a baseline and track activity against it, you can get baseline data and actual data, as well as visual indications of your progress.

You can use a baseline to debrief yourself or the team at any point in a project. The baseline is especially useful at the end of a phase, when you can compare actual progress against estimates from many weeks or months ago. You can then improve your planning skills and learn to make more accurate estimates up front. You can also use a baseline and the actual activity that you track against it to explain delays or cost overruns and to illustrate the impact of significant scope changes to employers or clients.

It's a good practice to revisit the baseline and estimates at the end of each phase of the project. Plus, you'll likely forget what you were thinking when you produced the original estimates.

Finally, you can also save and clear baselines for only selected tasks. If one task is thrown off track by a major change, for example, you can modify its estimates and leave the rest of the baseline alone.

## Saving a baseline

You can save a baseline at any time by opening the Set Baseline dialog box. After you save a baseline the first time, the baseline data of a summary task is updated if you make changes to a subtask below it, such as changing the start date for a task or deleting a task. However, when you're saving the baseline for only a selection of tasks, you can change that functionality by making a choice about how the baseline *rolls up*, or summarizes, data. You can choose to have modifications rolled up to all summary tasks or only from subtasks for any summary tasks you select. This second option works only if you've selected summary tasks and haven't selected their subtasks.

To use rolling wave planning, save a baseline for only selected tasks. In other words, perhaps you've sufficiently detailed the early tasks in the first phase or in the first month or two of a large project, and you're ready to commit to that part of the plan, but the latter detail still needs to be finalized. Saving a baseline for earlier tasks allows you and the team to get working on those tasks while allowing for the progressive detailing of later tasks down the line.

To save a baseline, follow these steps:

1. **To save a baseline for only certain tasks, select them by dragging over their task ID (row) numbers.**
2. **Click the Project tab, and in the Schedule group, click the Set Baseline drop-down arrow and choose Set Baseline.**

   The Set Baseline dialog box appears, with the Set Baseline radio button selected, as shown in Figure 14-1.
3. **Select either the Entire Project radio button or the Selected Tasks radio button.**
4. **If you chose Selected Tasks in Step 3, make selections in the Roll Up Baselines area.**

   You can summarize changed data in all summary tasks or for only selected summary tasks.
5. **Click the OK button to set the baseline.**

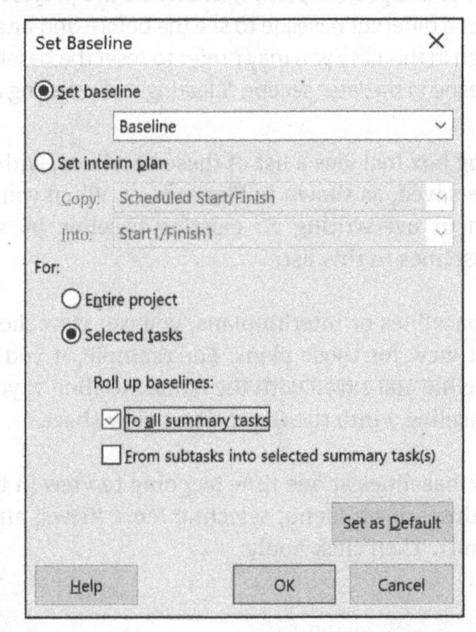


**FIGURE 14-1:** Setting baselines and interim plans.

TIP

Don't baseline the whole project if part of the plan involves iterative or incremental development or Agile methods; instead just baseline the parts of the plan that have fixed scope.

## Saving more than one baseline

The concept of multiple baselines seems almost contrary to the definition of a baseline. After all, the purpose of a baseline is to set the proverbial stake in the sand and measure progress against it. You might keep multiple baselines for several purposes. Consider these reasons:

- **You have a baseline plan for a customer and another one for the internal team.** In other words, you might want to steer the team to a tighter deadline than the promised delivery date.
- **You have multiple scenarios for the project.** Maybe you're unsure whether you'll need an extra widget for the gadget you're building. You can set one baseline that assumes the extra widget is unnecessary and set another baseline that assumes it's vital.
- **You're anticipating a risk event taking place.** You want to develop a recovery plan or response plan in one baseline version of the plan.
- **You're analyzing the effect of a change to the project schedule (or another element of the project plan) that affects the project or product scope.** You can set a different baseline to see the before-and-after effect of the change, though in this case you may prefer to reset the baseline. I discuss resetting the baseline in the later section "Clearing and resetting a baseline."

The Set Baseline dialog box includes a list of these baselines, with a date stamp of the last date each was saved, as shown in Figure 14-2. When you save a baseline, you can save it without overwriting an existing baseline by simply selecting another one of the baselines in this list.

If you save multiple baselines or interim plans, you can view them by displaying columns in any sheet view for those plans. For example, if you want to display information for a baseline you saved with the name Baseline 7, you need to insert the column named Baseline 7 into the Gantt Chart view sheet.

You can view multiple baselines at one time by going to View in the Task Ribbon, clicking to show the drop-down menu, selecting More Views, and then selecting Multiple Baselines Gantt. Then click Apply.

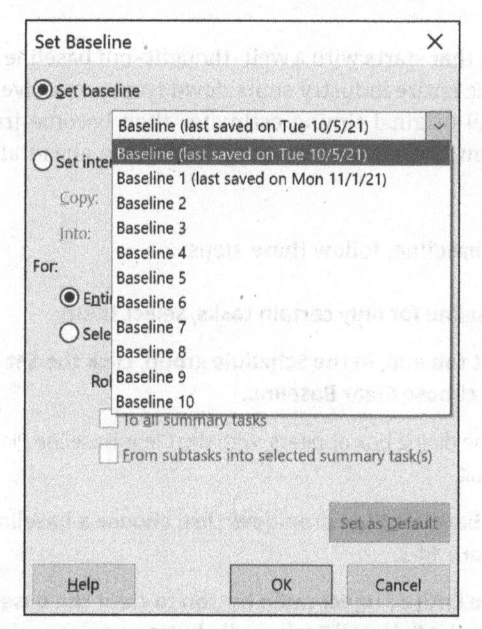


**FIGURE 14-2:** Multiple baseline options.

© John Wiley & Sons, Inc.

WARNING

When calculating variances (whether the schedule is early or late or costs are over or under budget), Project uses only the first baseline that's saved (the one named Baseline, not Baseline1). If the project schedule and other parameters change dramatically, such as when a project is delayed and later revived, you produce more valid variance data by clearing and resetting the baseline, as described in the next section.

WARNING

The more baselines you save, the larger the Project file becomes and the more it can slow performance.

## Clearing and resetting a baseline

A baseline is intended to be a frozen picture of the project plan that remains sacrosanct and never changes. Well, that's the theory. In practice, events can make an original baseline less than useful — and even obsolete.

For example, if a project that spans four years from beginning to end, you may want to save a new baseline every year because costs increase or resources change. Then you can see incremental versions of the estimates that reflect the changes you made based on changes in the real world.

In contrast, a project that starts with a well-thought-out baseline plan can change a week later when the entire industry shuts down from a massive strike that lasts for three months. All original timing estimates then become irrelevant, so you would adjust the plan, save a new baseline, and move ahead after the strike is resolved.

To clear an existing baseline, follow these steps:

1. **To clear the baseline for only certain tasks, select them.**
2. **Click the Project tab and, in the Schedule group, click the Set Baseline drop-down and choose Clear Baseline.**

   The Clear Baseline dialog box appears, with the Clear Baseline Plan option selected by default.
3. **From the Clear Baseline Plan drop-down list, choose a baseline to clear, as shown in Figure 14-3.**
4. **Select either the Entire Project radio button to clear the baseline for the entire project or the Selected Tasks radio button to clear selected tasks.**
5. **Click the OK button.**

   The project baseline is cleared or the selected tasks are cleared.

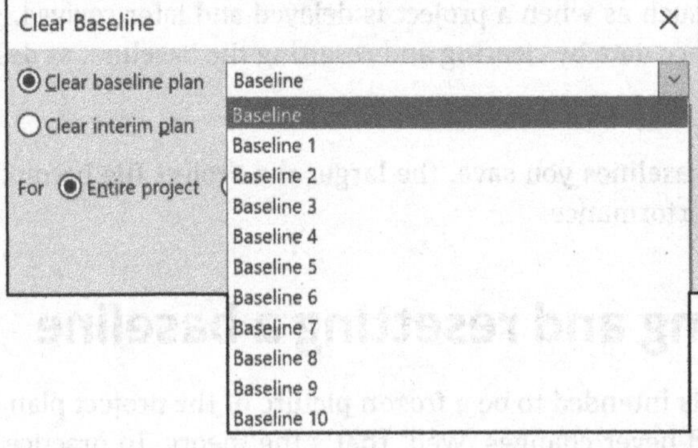


**FIGURE 14-3:** Clearing the baseline.

*© John Wiley & Sons, Inc.*

# In the Interim

An interim plan is sort of a "baseline light," in which you save only the actual start and finish dates of tasks that have had activity tracked on them, as well as the baseline start and finish dates for all tasks that haven't started.

An interim plan saves only timing information; other information, such as all the data about resource assignments and costs, is not saved. If timing information is all you need, an interim plan is a great way to save space. (When you save a large number of baselines, the Project file is *huge*. Saving an interim schedule makes the schedule less bulky.)

Because a baseline holds a great deal of data, the baseline can eventually become obsolete. An interim plan can be saved to record date changes but not overwrite the original baseline cost data.

Finally, you're allowed to save as many as 11 baseline plans (the original baseline plus baselines 1–10). If you need more than 11, consider using interim and some baseline data to expand the number of sets of data you can save.

TIP

Don't go crazy for baselines and interim plans. Even in a long-term project, saving too many sets of plans can become confusing. To keep track of a plan when you save it, print a copy for your records and note when and why the plan was saved.

## Saving an interim plan

Interim plans and baselines are saved using the same dialog box. The difference is that you have to specify where the data in an interim plan originates. For example, if you want to save the start and finish dates from the Baseline2 plan to the third interim plan, you copy from Baseline2 to Start/Finish3. If you want the current scheduled start and finish dates for all tasks, choose to copy from Start/Finish.

If you want to change currently scheduled start and finish dates in a baseline, but you don't want to change all the other data that's typically saved in a baseline, you can copy from Start/Finish to the baseline plan that you want to change.

Follow these steps to save an interim plan:

1. **To save an interim plan for only certain tasks, select them.**
2. **Click the Project tab and, in the Schedule group, click the Set Baseline drop-down arrow and choose Set Baseline.**

   The Set Baseline dialog box appears.
3. **Select the Set Interim Plan radio button.**
4. **From the Copy drop-down list box, select the set of data that you want to copy to the interim plan.**

5. **From the Into drop-down list box, select the fields in which you want to store the interim plan data, as shown in Figure 14-4.**
6. **Select the appropriate radio buttons to save the plan for the entire project or selected tasks.**
7. **Click the OK button to save the plan.**

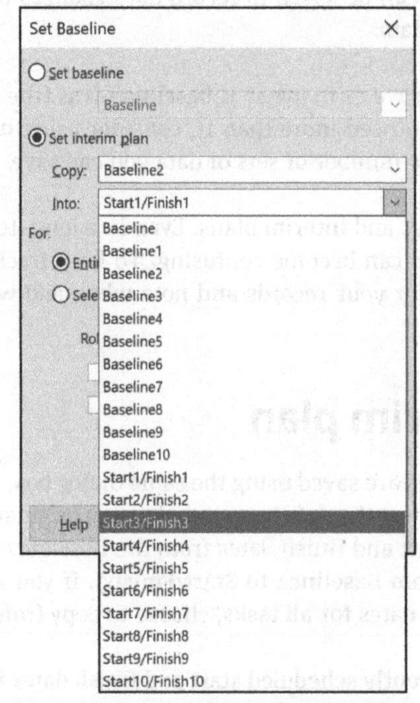


**FIGURE 14-4:** Copy settings from any saved baseline to an interim plan.



By using the Copy and Into fields in the Set Baseline dialog box, you can save as many as ten interim plans based on either baseline or actual data.

## Clearing and resetting an interim plan

Ten interim plans may seem like a lot when you're starting out in Project, but in the thick of a busy and ever-changing project, ten may fall short of the number you need. You may eventually need to clear one and resave it.

Project piggybacks baseline and interim plan settings, so you choose the Clear Baseline menu command to clear an interim plan.

To clear an interim plan, follow these steps:

1. **To clear only certain tasks in an interim plan, select them.**
2. **Click the Project tab and, in the Schedule group, click the Set Baseline drop-down arrow and choose Clear Baseline.**

   The Clear Baseline dialog box appears, as shown in Figure 14-5. You can use it to clear and reset interim plans as often as you like.
3. **Select the Clear Interim Plan radio button, and then, from the accompanying drop-down list, choose the plan that you want to clear.**
4. **Click an option button to specify whether to clear the specified interim plan for the entire project or for any tasks you've selected.**
5. **Click the OK button to clear the plan.**

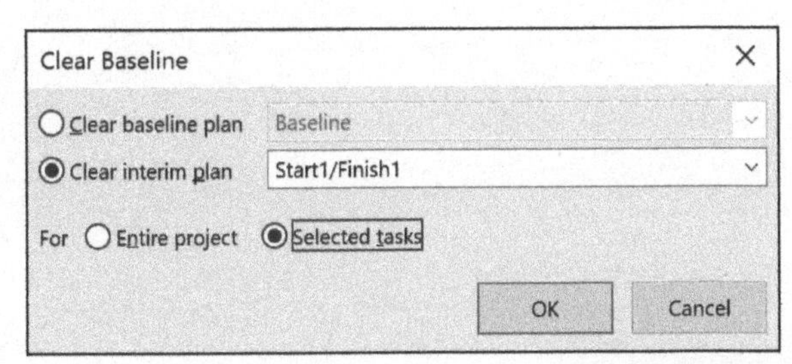


**FIGURE 14-5:** The Clear Baseline dialog box.



You can now save a new set of information to that interim plan, if you like.

WARNING

You may want to save a backup copy of various versions of the file with interim and baseline data. If you clear a baseline or an interim plan, its data disappears forever.

# 4

# Staying on Track

IN THIS PART . . .

Collect data to determine the project's status.

Compare planned accomplishments to actual accomplishments.

Figure out how to recover when a project doesn't proceed as planned.

Create reports to communicate project status.

IN THIS CHAPTER

» **Using tracking tools**

» **Recording activities on tasks**

» **Specifying the amount of work complete**

» **Updating fixed costs**

» **Tracking overtime hours**

» **Moving a task**

» **Using Update Project to make changes to the big picture**

# Chapter 15
# On the Right Track

**Project is a very powerful tool that can do amazing things with the data you provide. You can track all kinds of information, such as**

» Planned start and finish dates compared to actual start and finish dates

» Planned duration compared to actual duration

» Estimated costs and actual costs

» Percent complete

» Estimated remaining duration

» Hours worked by resource

» And more!

**In this chapter, I show you different ways to track your project's performance. In Chapter 16, you learn how to use some cool techniques to visualize project status.**

# Tracking Views

You already know that Project has a view for everything you want to do. To track project performance, you can use a lot of views, depending on the type of information you want and the amount of detail you need. For example, you can:

- » Use the default Task Sheet view and insert a % Complete column
- » Use the Task Sheet view with a Tracking table
- » Show the Task Usage view
- » Show the Resource Usage view and add a row to show actual work

I walk through how to access each of these views shortly and discuss how and when to use them, but first I want to show you how to set the status date.

## Setting the status date

Before you track anything, you need to tell Project the date for which you want to show status. By default, Project uses the date-and-time setting on your computer as the date for entering actual activity information. However, sometimes you want to time-travel. Suppose that the boss asks for a report showing the status of the project as of the last day of the quarter, September 30. You have gathered timesheets from all resources through that date, but you didn't enter those updates into Project until after the quarter ended. You can set the status date in Project to September 30 and then enter the tracking data.

Here's how to set the status date:

1. **Select the Project tab and then click the Status Date calendar icon in the Status group.**

   The Status Date dialog box appears.

2. **Click the down arrow to display the calendar.**
3. **If you want to set the status date in another month, click the right or left arrow in the upper-right corner of the calendar to navigate to that month.**
4. **Click the date you want.**
5. **Click the OK button.**

Now you're ready to start entering tracking data.

# Tracking status with the Task sheet

The simplest way to track status is to add a % Complete column to the Task sheet. Simply click the column header to the right of where you want to insert the column, and then right-click anywhere in the column. A list of options appears; select Insert Column. You then see a drop-down list with lots of options. Fortunately, the % Complete option is at the top of the list! Just select % Complete. Now you can enter a status for any tasks that have started.

This method of tracking status is the quickest and easiest. However, if you need information on resource usage or costs, this is not the right option.

# Using the Tracking table

If you like working with the Task sheet, you can level up your tracking game by using a Tracking table, as shown in Figure 15-1. Just follow these steps:

1. **Go to the View tab and make sure Gantt Chart view is selected in the Task Views group.**
2. **In the Data group, click the arrow under Tables.**
3. **Select Tracking.**

   In this view you can enter the actual start dates, finish dates, and duration of tasks. You can enter the percent complete or you can "Mark on Track" (the next section explains how to use Mark on Track). Project automatically calculates the remaining duration, actual work, and actual cost based on the percent complete.

| Task Name | Act. Start | Act. Finish | % Comp. | Phys. % Comp. | Act. Dur. | Rem. Dur. | Act. Cost | Act. Work |
|---|---|---|---|---|---|---|---|---|
| **Perimeter** | **Mon 1/3/22** | **NA** | **37%** | **0%** | **25.81 days** | **44.19 days** | **$10,650.00** | **426 hrs** |
| **Walls** | **Mon 1/3/22** | **NA** | **59%** | **0%** | **29.13 days** | **19.88 days** | **$10,650.00** | **426 hrs** |
| Dig trench | Mon 1/3/22 | Mon 1/17/22 | 100% | 0% | 2 wks | 0 wks | $3,000.00 | 120 hrs |
| Install footing | Mon 1/17/22 | Mon 1/24/22 | 100% | 0% | 1 wk | 0 wks | $2,000.00 | 80 hrs |
| Install cinder blocks | Mon 1/24/22 | NA | 94% | 0% | 2.83 wks | 0.18 wks | $5,650.00 | 226 hrs |
| Apply stucco | NA | NA | 0% | 0% | 0 wks | 1 wk | $0.00 | 0 hrs |
| Paint | NA | NA | 0% | 0% | 0 wks | 1 wk | $0.00 | 0 hrs |
| Apply finishes | NA | NA | 0% | 0% | 0 days | 4 days | $0.00 | 0 hrs |
| Wall contingency | NA | NA | 0% | 0% | 0 wks | 1 wk | $0.00 | 0 hrs |
| Walls complete | NA | NA | 0% | 0% | 0 days | 0 days | $0.00 | 0 hrs |
| **Entry gates** | **NA** | **NA** | **0%** | **0%** | **0 days** | **40 days** | **$0.00** | **0 hrs** |
| Perimeter Complete | NA | NA | 0% | 0% | 0 days | 0 days | $0.00 | 0 hrs |
| **Equipment** | **NA** | **NA** | **0%** | **0%** | **0 days** | **96 days** | **$0.00** | **0 hrs** |
| **Asset Management** | **NA** | **NA** | **0%** | **0%** | **0 days** | **47 days** | **$0.00** | **0 hrs** |
| **Operations Readiness** | **NA** | **NA** | **0%** | **0%** | **0 days** | **40 days** | **$0.00** | **0 hrs** |

**FIGURE 15-1:** The Task sheet with the Tracking table.

REMEMBER

If you mark that a task is complete, and you don't enter the actual start or finish dates, Project (ever the optimist) assumes that you started and finished according to plan. If your start and end dates are different, you just enter the actual dates in the appropriate columns by using the drop-down calendar in the Actual Start or Actual Finish column. However, if you don't modify the task duration, and you enter an earlier finish date, be aware that Project moves the start date earlier by a corresponding amount.

## Tracking buttons

Project has some handy little tracking buttons in the Schedule group of the Task tab. As Figure 15-2 shows, you can click a button to mark tasks 0%, 25%, 50%, 75%, and 100% complete. For the most part, this is all you need. Most people don't care if a task is 15 or 20 percent complete. They want a general idea of the status. These buttons are an easy way to show status without haggling over how far along the work is. I use the following guidelines:

- If a task hasn't started, it is 0 percent complete (obviously).
- If a task has started, but is less than half done, it is 25 percent complete.
- A task that is approximately half done is 50 percent complete.
- A task that is significantly more than half done is 75 percent complete.
- And of course, a task that is done is 100 percent complete.

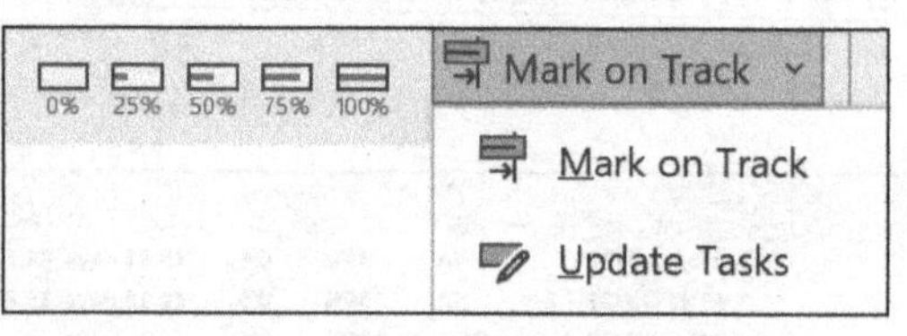


**FIGURE 15-2:** The tracking buttons.

© John Wiley & Sons, Inc.

Another useful shortcut is using the Mark on Track button. Use this button for tasks that are on time. Project calculates the percent of work done based on the duration and the date. Figure 15-3 shows the project information using the Mark on Track button.

| Task Name | Act. Start | Act. Finish | % Comp. | Phys. % Comp. | Act. Dur. | Rem. Dur. | Act. Cost | Act. Work |
|---|---|---|---|---|---|---|---|---|
| Perimeter | Mon 1/3/22 | NA | 27% | 0% | 18.72 days | 51.28 days | $7,450.00 | 298 hrs |
| Walls | Mon 1/3/22 | NA | 43% | 0% | 21.13 days | 27.88 days | $7,450.00 | 298 hrs |
| Dig trench | Mon 1/3/22 | Mon 1/17/22 | 100% | 100% | 2 wks | 0 wks | $3,000.00 | 120 hrs |
| Install footing | Mon 1/17/22 | Mon 1/24/22 | 100% | 100% | 1 wk | 0 wks | $2,000.00 | 80 hrs |
| Install cinder blocks | Mon 1/24/22 | NA | 41% | 50% | 1.23 wks | 1.78 wks | $2,450.00 | 98 hrs |

**FIGURE 15-3:** Tracking table with the Mark on Track button.



## Determining the percent complete

If using the tracking buttons isn't a good option for your project, you can also calculate the percent complete in more precise ways. For example, if you estimate that a task should take ten hours of effort, and your resources report performing five hours of effort, you can say that the task is 50 percent complete. Be careful, though: Just because people have spent half the allocated time doesn't mean that they have accomplished half the work.

Another way to estimate the percent complete is by cost. If you originally estimated that four resources assigned to a four-day task would tally $4,000 in costs, and the total time that the resources report spending on the task is $3,000, you can estimate that the task is 75 percent finished. But again, just because you have spent three quarters of the money doesn't mean that you have accomplished three quarters of what you set out to do. The best rule of thumb is trust and verify.

Determining the percentage of work that's done is easier when the deliverable is measurable. For example, if your task is to install updates on 100 workstations in four days, and you've installed the updates on 25 workstations, the task is 25 percent finished.

But not every task can be calculated as neatly. If you know it will be hard to gauge the percent complete when you're in the planning phase, collaborate with the team member on the interim measurements to allocate 25, 50, 75, and 100 percent complete. For example, if you want to determine interim measurements for the three-week task titled Develop Medical Protocols, you might agree that when a complete list of protocols is needed and they've been assigned to development, the task is 25 percent complete. When 50 percent is drafted, the task is 50 percent complete. When the protocols are 100 percent drafted, the task is 75 percent complete. When the list has been edited and someone in charge has signed off on it and approved it, the task is 100 percent complete.

REMEMBER

To avoid finding out (usually at the last minute) that a team member's version of "almost complete" is significantly different from yours, agree up front that you'll sign off on 75 percent complete only after specific criteria are met.

# Tracking status with Task Usage view

The Task Usage view lets you record detailed resource information for each task. For example, let's say Luis and Bruce are both working on applying stucco. This is an 80-hour task. Project assumes that Luis and Bruce will split the work equally — 40 hours each. They are both scheduled to work eight hours per day Tuesday through Friday. However, Bruce ends up only working four hours on Wednesday and Thursday. This impacts the duration and potentially the budget as well, especially if your resources have different rates. This view lets you record the actual hours each resource worked.

You can access the Task Usage view on the View tab, in the Task Views group, as shown in Figure 15-4. To enter the actual work rows for this view, right-click anywhere on the chart and select Actual Work.

| Task Mode | Task Name | Work | Duration | Start | Finish | Details | M | T | W | T | F | S | S | M | T |
|---|---|---|---|---|---|---|---|---|---|---|---|---|---|---|---|
| | Apply stucco | 80 hrs | 1.2 wks | Mon 2/14/22 | Tue 2/22/22 | Work | 2h | 16h | 12h | 12h | 16h | | | 15h | 7h |
| | | | | | | Act. Wo | 2h | 16h | 12h | 12h | 16h | | | 15h | |
| | Luis Pena | 40 hrs | | Mon 2/14/22 | Mon 2/21/22 | Work | 1h | 8h | 8h | 8h | 8h | | | 7h | |
| | | | | | | Act. Wo | 1h | 8h | 8h | 8h | 8h | | | 7h | |
| | Bruce Hashimoto | 40 hrs | | Mon 2/14/22 | Tue 2/22/22 | Work | 1h | 8h | 4h | 4h | 8h | | | 8h | 7h |
| | | | | | | Act. Wo | 1h | 8h | 4h | 4h | 8h | | | 8h | |

**FIGURE 15-4:** Task Usage view.



When there is a variance in the actual work hours compared to the planned work or the baseline work, Project automatically adjusts the duration and the end date. In Figure 15-4, the Apply Stucco task originally had a duration of 1 week but now it shows a duration of 1.2 weeks and shows hours for Bruce on Tuesday of the following week. There is also a bar chart with a pencil in the Indicator column that informs you there were edits made to the task.

You can enter only the total hours by inserting a column and selecting Actual Work, then entering the total number of hours for the task or resource. If you do this, Project adds or deletes hours on a daily basis. It adds and subtracts hours starting from the last scheduled day of work for a task.

Another option is to enter the hours each resource spent on a daily basis. To do this, add the Actual Work row by right-clicking the chart and choosing Actual Work. Then enter the actual hours worked by each resource. Project updates the Work Column to indicate the Work by Resource and the Work for the Task.

# Tracking status with Resource Usage view

Resource Usage view, as shown in Figure 15-5, lets you see all the tasks each resource is working on. This view also lets you see how many hours a person is committed to the project. This can be handy if you are paying someone hourly.

Like with the Task Usage view, you can add a row for actual work. If the resource is working on tasks back-to-back, and one task takes longer, Project adjusts the start of the following tasks.

| Resource Name | Work | Details | M | T | W | T | F |
|---|---|---|---|---|---|---|---|
| Ty Mack | 280 hrs | Work | 8h | 8h | 8h | 8h | 8h |
| | | Act. Wo | | | | | |
| *Identify requirements* | *120 hrs* | Work | 8h | 8h | 8h | 8h | 8h |
| | | Act. Wo | | | | | |
| *Alternatives analysis* | *120 hrs* | Work | | | | | |
| | | Act. Wo | | | | | |
| *Source Selection* | *40 hrs* | Work | | | | | |
| | | Act. Wo | | | | | |

**FIGURE 15-5:** Resource Usage view.



## Uh-oh — you're in overtime

When you enter 16 hours of work in a single day for a resource, even though the resource's availability is based on a calendar with an 8-hour day, Project doesn't recognize those hours as overtime. It's one instance of having to lead Project by the hand and tell it to specify overtime work.

When you enter hours in the Overtime Work field, Project interprets them as the number of total work hours that are overtime hours. So if you enter **16** hours of work on a task in the Work column and then enter **4** in the Overtime Work column, Project assumes 12 hours of work at the standard resource rate and 4 hours at the overtime rate.

To enter overtime hours, follow these steps:

1. **Display Resource Usage view.**
2. **Right-click a column heading and choose Insert Column.**

   The list of columns appears.
3. **Select Overtime Work in the list.**
4. **Click in the Overtime Work column for a specific assignment for a resource and then enter the overtime hours.**

Figure 15-6 shows that you have allocated 4 hours of overtime to Luis Pena.

| Resource Name | Work | Overtime Work |
|---|---|---|
| Luis Pena | 352 hrs | 4 hrs |
| *Dig trench* | *80 hrs* | *0 hrs* |
| *Install footing* | *40 hrs* | *4 hrs* |
| *Install cinder blocks* | *120 hrs* | *0 hrs* |
| *Apply stucco* | *40 hrs* | *0 hrs* |
| *Paint* | *40 hrs* | *0 hrs* |
| *Apply finishes* | *32 hrs* | *0 hrs* |

**FIGURE 15-6:** Assigning overtime.



If you specify overtime, Project assumes that *effort-driven* tasks have a shorter duration (for a refresher on effort-driven tasks, see Chapter 5). After all, if the task were to take three 8-hour days (24 hours of work) to complete, and you recorded the resource working 12 hours for two days in a row, Project figures that all the effort was accomplished in less time. The duration of the task *shrinks*. If that's not what happened, you have to manually modify the task duration.

## Specifying remaining durations for auto-scheduled tasks

Tracking information for auto-scheduled tasks has a weird and wonderful relationship in Project. For example, Project tries to help you by calculating the duration based on other entries you make, such as actual start dates and finish dates. In that particular case, Project calculates task duration according to those dates. (This process works in reverse, too: If you enter the task duration, Project recalculates the finish date to reflect it.)

You may instead want to enter the duration yourself. For example, if you have entered a start date and four weeks of work on a task that has a baseline of three weeks of work, Project can't comprehend that the scope of the task has changed, and that now the task is incomplete and requires another week of work. You have to tell Project about it.

To modify the duration of a task (either in progress or completed), follow these steps:

1. **Display Gantt Chart view.**
2. **Click View, and in the Data group, click Tables and then click Tracking.**

   The Tracking table is displayed, as shown in Figure 15-7.
3. **Click in the Actual Duration column of the task you want to modify and then adjust the actual duration up or down using the arrows. Or just enter the actual duration.**

4. **If you want to enter a remaining duration, click in the Rem. Dur. (Remaining Duration) column and type a number and an increment symbol (such as d for days or w for weeks).**

| Task Name | Act. Start | Act. Finish | % Comp. | Phys. % Comp. | Act. Dur. | Rem. Dur. | Act. Cost | Act. Work |
|---|---|---|---|---|---|---|---|---|
| Perimeter | Mon 1/3/22 | NA | 28% | 0% | 19.69 days | 49.44 days | $8,100.00 | 320 hrs |
| Walls | Mon 1/3/22 | NA | 46% | 0% | 22.56 days | 26.57 days | $8,100.00 | 320 hrs |
| Dig trench | Mon 1/3/22 | Mon 1/17/22 | 100% | 100% | 2 wks | 0 wks | $3,000.00 | 120 hrs |
| Install footing | Mon 1/17/22 | Tue 1/25/22 | 100% | 100% | 1 wk | 0 wks | $2,100.00 | 80 hrs |
| Install cinder blocks | Mon 1/24/22 | NA | 50% | 50% | 1.5 wks | 1.5 wks | $3,000.00 | 120 hrs |
| Apply stucco | NA | NA | 0% | 0% | 0 wks | 1 wk | $0.00 | 0 hrs |
| Paint | NA | NA | 0% | 0% | 0 wks | 1 wk | $0.00 | 0 hrs |
| Apply finishes | NA | NA | 0% | 0% | 0 days | 4 days | $0.00 | 0 hrs |
| Wall contingency | NA | NA | 0% | 0% | 0 wks | 1 wk | $0.00 | 0 hrs |
| Walls complete | NA | NA | 0% | 0% | 0 days | 0 days | $0.00 | 0 hrs |



**FIGURE 15-7:** Gantt chart with Tracking table.

WARNING

If you enter the percent complete for a task and then modify the duration to differ from the baseline, Project automatically recalculates the percent complete to reflect the new duration. For example, if you enter 50 percent complete on a 10-hour task and then modify the actual duration to 20 hours, Project considers those 5 hours (50 percent of 10 hours) being only 25 percent of the 20 hours.

## Entering fixed-cost updates

*Fixed costs* are costs that aren't influenced by time, such as equipment purchases and consulting fees. Compared with the calculations and interactions of the percent completes and the start and finish dates for hourly resources, fixed-cost tracking will seem like simplicity itself!

Here's how to track fixed costs:

1. **Display Gantt Chart view.**
2. **Click View, and in the Data group, click Tables and click Cost.**

   The Cost table is displayed. Click the Add New Column header and select Fixed Costs from the drop-down options.
3. **Click in the Fixed Cost column for the task you want to update.**
4. **Type the fixed cost, or a total of several fixed costs, for the task.**

That's it! However, because Project lets you enter only one fixed-cost amount per task, consider adding a note to the task to itemize the costs you've included in the total.

REMEMBER

The fixed cost assigned to the task is only the scheduled cost for the task. For this cost to become the planned cost, the task has to be baselined. Also, updates to the % Complete column update actual costs, depending on the accrual method that's selected for the task: Prorated is the default setting, but the start and finish dates of the task also influence the actual costs when the percent complete is applied.

TIP

Consider using some of the ten customizable Cost columns for itemized fixed-cost entry. Rename one Equipment Purchase, and another Facility Rental, for example, and then enter the costs in those columns. Of course, these columns of data may not perform calculations such as rolling up total costs to the summary tasks in the project, but they serve as reminders about itemized fixed costs.

REMEMBER

In Project, you can designate a resource type as a cost and assign a cost every time you assign the resource. The Cost column calculation of total cost includes the amount spent for cost resources on tasks. See Chapter 7 for more about resource types.

# Moving a Task

If you've ever managed the subcontractors on any type of construction project, you've probably heard these dreaded words: "My current job is running long, so I'll get to your job in about a week." What you'd rather hear from a subcontractor (or any other type of resource) is this: "My schedule opened up, so I can start work on this assignment a week early." Project helps you easily reschedule an entire task or reschedule part of a task as of the status date.

Here's how to use the Move Task tool to move a task:

1. **Display Gantt Chart view.**
2. **If the task is partially complete and you want to move its uncompleted portion, select the Project tab. In the Status group, click the Status Date calendar icon and set a status date in the Status Date dialog box.**
3. **Select the task to move.**
4. **Click the Move button in the Tasks group on the Task tab.**

   The choices for rescheduling the task appear, as shown in Figure 15-8.

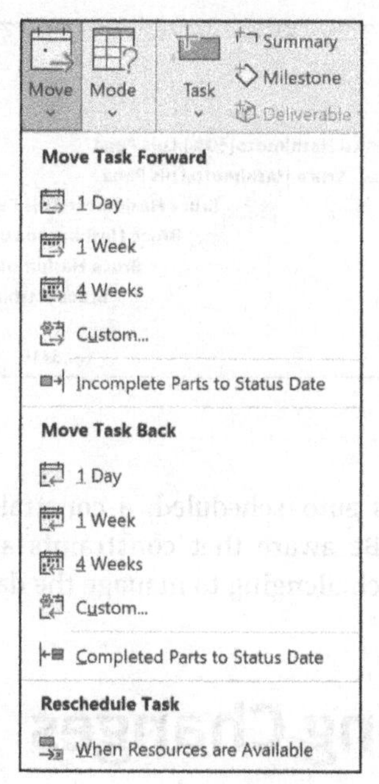


**FIGURE 15-8:** The Move dialog box.

*© John Wiley & Sons, Inc.*

5. **Specify how to move the task:**
   - *To move the task forward or back by a default amount:* Click 1 Day, 1 Week, or 4 Weeks in either the Move Task Forward or Move Task Back section.
   - *To move the task by a custom time frame:* Click the appropriate Custom option, specify the number of working days by which to move the task in the dialog box that appears, and click the OK button.
   - *To reschedule part of the task according to the status date:* Choose the Incomplete Parts to Status Date option to split the task and move the uncompleted portion to resume later, starting on the status date. Choose the Completed Parts to Status Date option to split the task and move the completed portion earlier, before the status date. Figure 15-9 shows that the Install Cinder Blocks task has now been split.
   - *To reschedule the task based on the availability of the assigned resources:* Choose the appropriate Custom option, specify the number of working days by which to move the task in the dialog box that appears, and click OK.

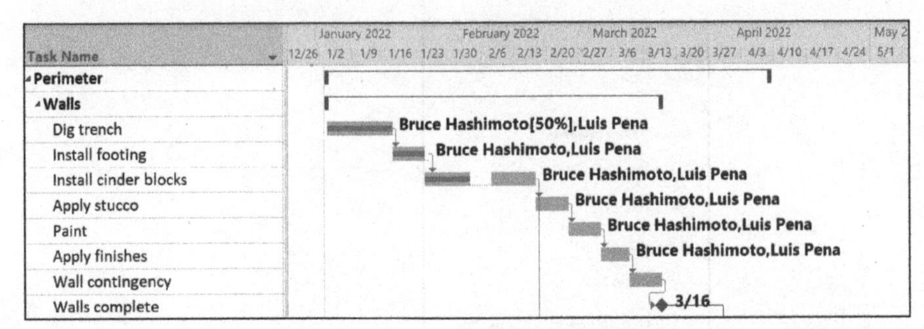


**FIGURE 15-9:** Moving a task.



When you're moving a task that's auto-scheduled, a constraint is applied, to honor the newly scheduled task. Be aware that constraints apply restrictions to task scheduling and can make it challenging to manage the dates.

# Update Project: Sweeping Changes

If it's been a while since you tracked activity, and you want to update the schedule, Update Project may be for you. This feature lets you track chunks of activity for specified periods. Update Project works best, however, if most tasks have been completed on schedule.

This type of tracking isn't finely tuned. You are basically trusting that Project can accurately depict your project status based on the current date.

Here are the settings that Update Project offers you:

- **Update Work As Complete Through:** You can update the project in one of two ways through the status date you specify in this box. The Set 0% – 100% Complete setting lets Project calculate the percent complete on every task that should have begun by that time. By choosing this option, you tell Project to assume that the tasks started and progressed exactly on time. The Set 0% or 100% Complete Only setting works a little differently: It tells Project to record 100 percent complete on tasks that the baseline indicated would be complete by now and to leave all other tasks at 0 percent complete.
- **Reschedule Uncompleted Work to Start After:** This setting reschedules the portions of tasks that aren't yet complete to start after the specified date.

To use Update Project, follow these steps:

1. **Display Gantt Chart view.**
2. **To update only certain tasks, select them.**
3. **On the Project tab, in the Status group, choose Update Project.**

   The Update Project dialog box appears, as shown in Figure 15-10.

   You can choose Update Work as Complete Through or Reschedule Uncompleted Work to Start After.

4. **If you click Update Work as Complete Through, choose the update method you prefer: Set 0% – 100% Complete or Set 0% or 100% Complete Only.**
5. **If you want a status date other than the one shown or you didn't previously set one, set the date to use in the field in the upper-right corner.**
6. **If you want Project to reschedule any work rather than update work as complete, select the Reschedule Uncompleted Work to Start After radio button and then select a date from the list.**
7. **Choose whether you want these changes to apply to the entire project or to only selected tasks.**
8. **Click the OK button to save the settings and have Project make updates.**

If you want, you can use Update Project to make certain global changes, such as to mark as 100 percent complete all tasks that should be complete according to the baseline. Then you perform more detailed task-by-task tracking on individual tasks that are only partially complete.

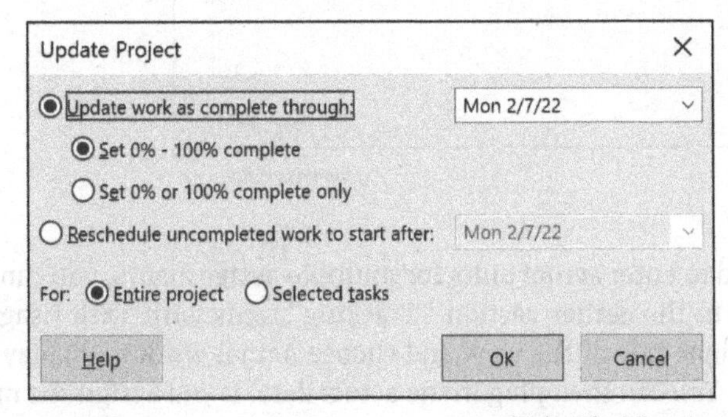


**FIGURE 15-10:** Using the Update Project feature.

# Tracking Materials

Tracking the amount of materials used on tasks involves tracking actual units at the level of material resources. If you create a resource named Rubber and assign it to the Manufacture Tires task at 500 tons, for example, and then you use only 450 tons, you enter the actual units used.

This situation resembles the way you track work resource hours on tasks. To make it happen, simply follow these steps:

1. **Display Resource Usage view.**
2. **Locate the material resource in the list, and under the resource name, double-click the assignment for which you want to enter the actual units.**

   The Assignment Information dialog box appears (see Figure 15-11).
3. **Click the Tracking tab.**
4. **Enter the actual units used in the Actual Work field.**
5. **Click the OK button.**

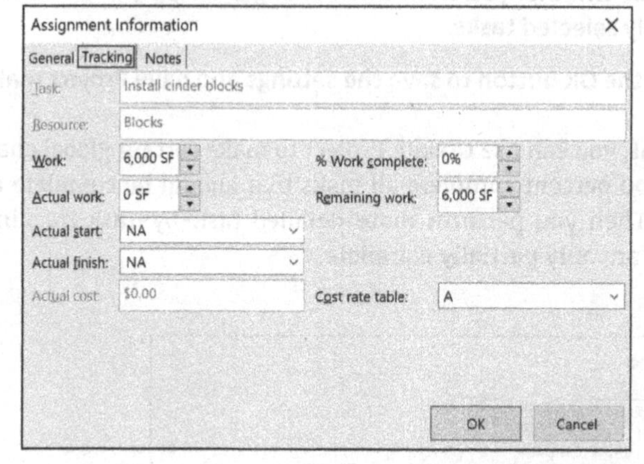


**FIGURE 15-11:** Tracking materials.

© John Wiley & Sons, Inc.

If you need to enter actual units for multiple assignments, you can use the method I describe in the earlier section "Tracking status with Task Usage view." Right-click the right side of the view and choose Actual Work to display the row labeled Act. Work. You can then plug in the actual data. If you assign the material resource of 500 tons to a five-day task, for example, remember that Resource Usage view shows Project spreading out the usage — 100 tons for each of the five days of the task. If you don't care which day the materials were used, you can simply increase or decrease one setting to make up for any difference in actual usage.

# Tracking More than One Project

In Chapter 2, in the section "Inserting one project into another," I describe how to show multiple projects, or subprojects, all in one project file. The inserted project is shown as a summary task with all subtasks hidden. To display all tasks in the inserted project, simply click the expansion triangle to the left of the summary task. This allows you to create dependencies between inserted projects in the consolidated file. If you have one project that can't start until another one finishes, for example, you can clearly see in the consolidated file how various separate projects in your organization affect each other.

If you check the Link to Project check box shown in Figure 15-12, any changes you make to one project are represented and updated in the consolidated file. If you don't link the files, changes in source files aren't reflected in consolidated files, and the consolidated file information doesn't affect the inserted project's information. You may create this type of unlinked consolidated file if you simply want to see a snapshot of how all projects are progressing and don't want to run the risk of your settings changing the source information.

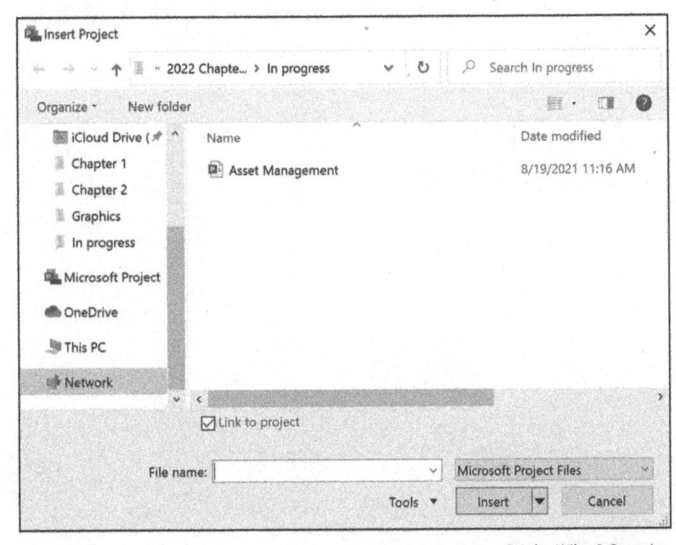


**FIGURE 15-12:** The Insert Project dialog box.

TIP

After you insert projects in a file, you can move them around by using the Cut and Paste tools. When you do this, message windows may appear, offering you options to resolve any conflicts that occur because of dependency links you created between inserted projects.

WARNING

If you want to insert multiple subprojects, leave them all collapsed. This strategy ensures that they're all inserted at the topmost level of the outline in the consolidated project.

IN THIS CHAPTER

» Viewing your progress with indicators and taskbars

» Seeing progress from various Project views

» Examining cost and time variances

» Understanding earned value

» Observing multiple critical paths

# Chapter 16

# Project Views: Observing Progress

Some project managers use Project simply to paint a picture of all the effort their project will entail and then stash the plan in a drawer. That's a mistake. After you enter all the project data, save a baseline, and then track actual activity on the project, you receive in return an amazing array of information from Project that can help you stay on time and on budget.

Once you track some activity on several tasks, Project lets you view baseline estimates alongside the real-time plan. Project alerts you to tasks that are running late and also shows how the critical path shifts over time.

Project also provides detailed performance information. In fact, the information you can see about the cost and schedule performance can be quite robust. You can produce a simple cost-and-schedule variance, or you can use more complex metrics, such as a cost performance index using earned value management techniques. I discuss this particular measurement technique in the later section "Tracking Progress Using Earned Value Management."

Keep the project file close at hand — and take a look at how Project can make you the most informed project manager in town.

## Seeing Where Tasks Stand

You've diligently entered resource work hours on tasks, recorded the progress on tasks, and entered fixed costs. Now what? Well, all that information has initiated several calculations and updates to the project schedule. It's time to take a quick look at the changes that all this tracking has produced in the project schedule.

TIP

A check mark in the Indicator field tells you that a task is complete. If you see an unfamiliar indicator icon crop up in the project, hold the mouse pointer over the indicator. A text box opens and describes its meaning. To see a list of Project icons and their meanings, click the Help button on the Help tab and type **indicators** in the Search box. You see a Help topic that says "Add Indicators." If you click that link, you see a hyperlink for the Indicators field. Clicking that link shows you the indicator icons and a description of the last several editions of Project.

### Baseline versus actual progress

One obvious way to view the difference between a baseline estimate and the progress you've tracked in the project is by using a taskbar. After you track progress on several tasks, the Gantt chart shows a dark bar superimposed on the baseline taskbar. For example, in Figure 16-1, Tasks 3 through 6 are complete; you can tell by the solid dark bar that extends the full length of the taskbar. Task 7 is only partially complete; the dark line that indicates actual progress only partially fills the duration for the task. Tasks 8 and 9 have no recorded activity on them; you see no dark line indicating progress — only the normal taskbar.

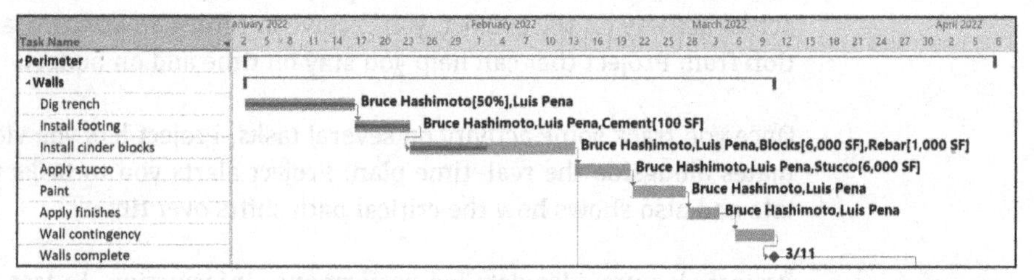


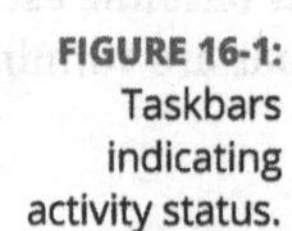

**FIGURE 16-1:** Taskbars indicating activity status.

© John Wiley & Sons, Inc.

REMEMBER

To track progress against a baseline, you have to save one. I cover setting a baseline in Chapter 14.

### Lines of progress

*Progress lines* offer additional visual indicators of how you're doing. As you can see in Figure 16-2, a progress line shows the progress as of a specified date. Activities that are behind are indicated when the line points to the left.

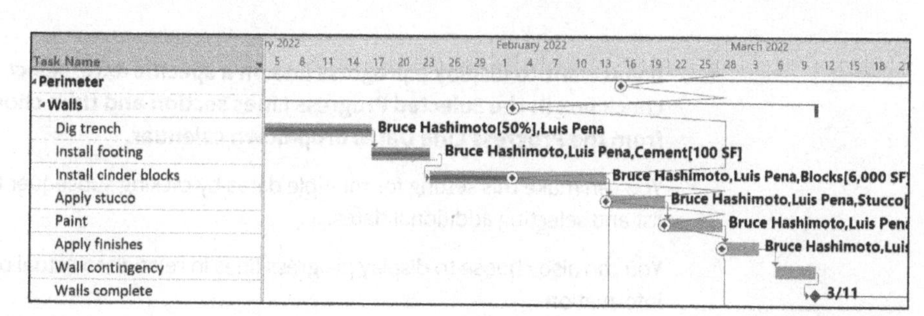


**FIGURE 16-2:** Progress lines, seemingly run amok.

© John Wiley & Sons, Inc.

In Figure 16-2, progress is measured as of the project status date. You can see that the tasks for Install Cinder Blocks, Apply Stucco, and Paint are behind, because the line points to the left.

## Displaying progress lines

By default, Project doesn't display progress lines. You have to turn them on. And while you're at it, you might as well specify when and how they appear. Here's how to display progress lines and change their settings:

1. **Display Gantt Chart view.**
2. **Right-click the chart portion of the view and choose Progress Lines.**

   The Progress Lines dialog box appears, as shown in Figure 16-3.
3. **If you want Project to always show a progress line for the current or status date, select the Display check box in the Current Progress Line section and then select the At Project Status Date radio button or the At Current Date radio button.**
4. **If you want progress lines to be displayed at set intervals, do this:**

   (a) *Select the Display Progress Lines check box in the Recurring Intervals section and then select the Daily radio button, Weekly radio button, or Monthly radio button.*

   (b) *Specify the interval settings.*

   For example, if you select Weekly, you can choose every week, every other week, and so on, as well as for which day of the week the line should be displayed on the timescale.
5. **Choose whether you want to display progress lines beginning at the Project start date or on another date.**

   To use the Project start date, simply select the Project Start radio button in the Begin At section of the dialog box. To select an alternative start date, select the second radio button and then select a date from the calendar drop-down list.

6. **If you want to display a progress line on a specific date, select the Display check box in the Selected Progress Lines section and then choose a date from the Progress Line Dates drop-down calendar.**

   You can make this setting for multiple dates by clicking subsequent lines in this list and selecting additional dates.

   You can also choose to display progress lines in relation to actual or baseline information.

   If a task has been tracked to show 50 percent complete, for example, and you choose to have Project display progress lines based on actual information, the peak appears relative to the 50 percent *actual* line, not to the complete baseline taskbar.

7. **Click the OK button to save these settings.**

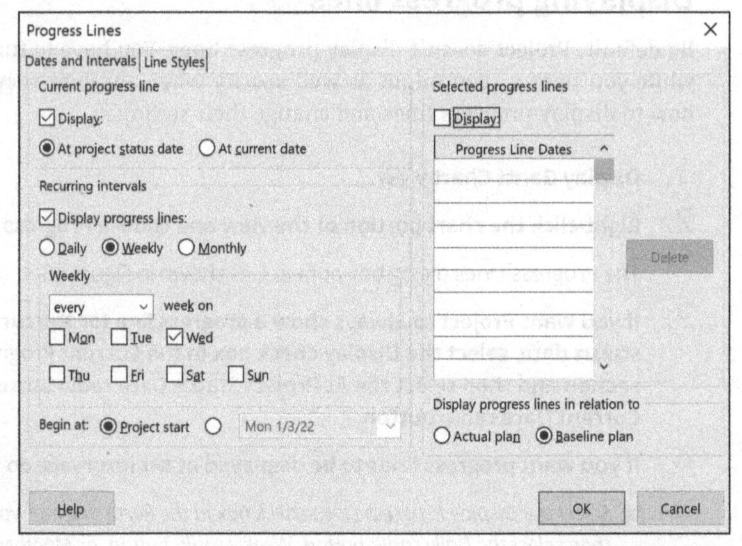


**FIGURE 16-3:** The Progress Lines dialog box.



To hide the progress lines, right-click anywhere on the chart and select Progress Lines from the context menu. In the Progress Lines dialog box, uncheck the Display check box and then click OK.

## Formatting progress lines

In keeping with the almost mind-boggling array of formatting options that Project makes available to you, you can modify how progress lines are formatted.

As with any changes to formatting, you're tampering with the way Project codes visual information for readers. Be cautious about making formatting changes that cause the plan to be difficult to read for those who are accustomed to the default Project formatting.

To modify progress-line formatting:

1. **Display Gantt Chart view.**
2. **Right-click the chart portion of the view and choose Progress Lines.**

   The Progress Lines dialog box appears (refer to Figure 16-3).
3. **Select the Line Styles tab.**

   You see the display options shown in Figure 16-4.
4. **In the Progress Line Type area, select a line style sample.**
5. **In the Line Type drop-down lists, select a style from the samples.**

   You can make two settings: one for the current progress line and one for all other progress lines.
6. **You can also change the line color, progress point shape, and progress point color by making different choices in those drop-down lists.**
7. **If you want a date to appear alongside each progress line, select the Show Date for Each Progress Line check box and then select a date from the Format drop-down list.**

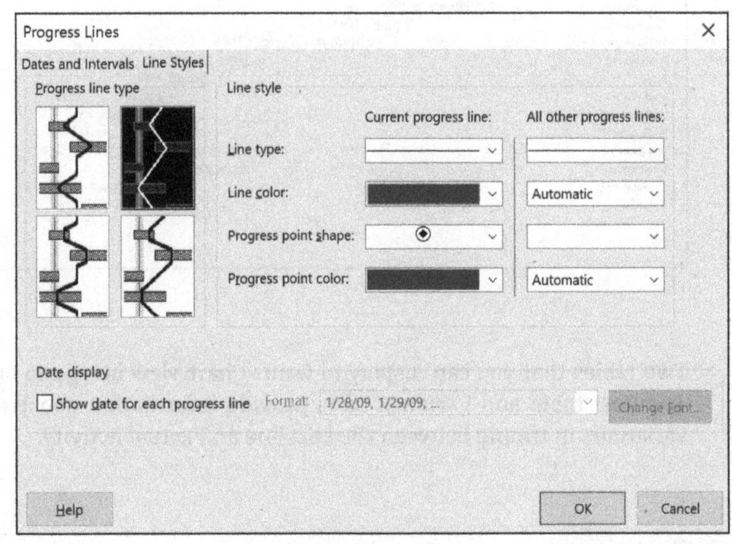


**FIGURE 16-4:** Progress line styles.

© John Wiley & Sons, Inc.

8. **If you want to change the font that's used for the displayed date, click the Change Font button and make the changes. Click OK in the Font dialog box.**
9. **Click the OK button to save these settings.**

# Delving into the Detail

Visual indicators such as taskbars and indicator icons are useful to alert you to delays or variances between estimated and actual performance, but they don't provide detailed information. To get the lowdown on how far ahead (or behind) you are, down to the day or penny, scan the numbers. The numbers that Project provides reveal much about whether you're on schedule and within the budget.

One way to view variances is to show Tracking Gantt view. From the Task tab, go to the View section on the far left side and click the down arrow. Choose Tracking Gantt. The summary task has a gray bar within it showing the status of the subtasks beneath it that are the farthest along. The subtasks indicate the progress by darkening the bar for the task to reflect the current status. As shown in Figure 16-5, in Tracking Gantt view, you can show exactly how much progress has been made on each task and subtask in the project.

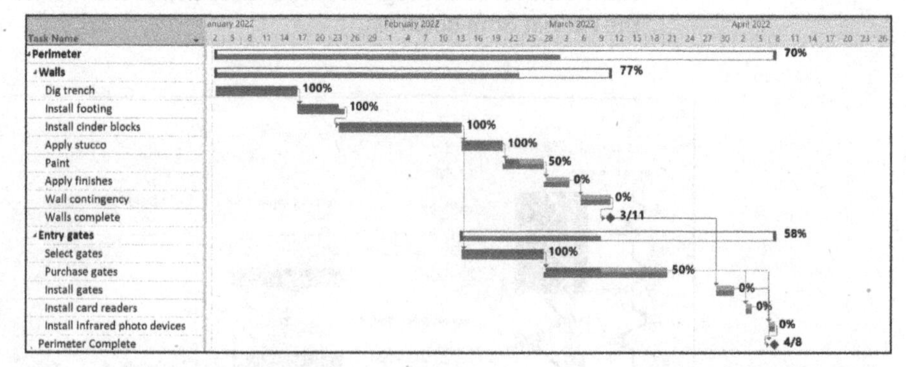


**FIGURE 16-5:** Tracking Gantt view.



Two tables that you can display in Gantt Chart view bring the options into focus. The Cost table and Variance table provide information about dollars spent and variations in timing between the baseline and actual activity.

REMEMBER

To display a table, choose View⇨Data⇨Tables and click the table name. In the Tracking Gantt view, you can also place the pointer on the blank square immediately above row 1 and to the left of the first column. Then right-click the mouse to display the available tables for easy selection.

In the Cost table shown in Figure 16-6, you can review data that compares baseline estimates to actual costs and the variance between them. For the Cost table, a variance that's positive indicates that you've spent more than expected. A variance that's negative indicates that you're under budget.

In the project shown in Figure 16-6, you can see that the task to dig the trench was $300 over the baseline cost. Installing the cinder blocks came in $7,200 under baseline, and applying the stucco was $2,375 over baseline. By looking at the detail, you can identify the exact activities that caused the variance and how much each activity contributed to the total variance.

| Task Name | Fixed Cost | Fixed Cost Accrual | Total Cost | Baseline | Variance | Actual | Remaining |
|---|---|---|---|---|---|---|---|
| **Perimeter** | **$0.00** | **Prorated** | **$236,250.00** | **$240,775.00** | **($4,525.00)** | **$217,610.00** | **$18,640.00** |
| **Walls** | **$0.00** | **Prorated** | **$213,125.00** | **$217,650.00** | **($4,525.00)** | **$210,475.00** | **$2,650.00** |
| Dig trench | $300.00 | Prorated | $3,300.00 | $3,000.00 | $300.00 | $3,300.00 | $0.00 |
| Install footing | $0.00 | Prorated | $2,700.00 | $2,700.00 | $0.00 | $2,700.00 | $0.00 |
| Install cinder blocks | ($7,200.00) | Prorated | $151,000.00 | $158,200.00 | ($7,200.00) | $151,000.00 | $0.00 |
| Apply stucco | $2,375.00 | Prorated | $52,500.00 | $50,125.00 | $2,375.00 | $52,500.00 | $0.00 |
| Paint | $0.00 | Prorated | $2,000.00 | $2,000.00 | $0.00 | $975.00 | $1,025.00 |
| Apply finishes | $0.00 | Prorated | $1,625.00 | $1,625.00 | $0.00 | $0.00 | $1,625.00 |
| Wall contingency | $0.00 | Prorated | $0.00 | $0.00 | $0.00 | $0.00 | $0.00 |
| Walls complete | $0.00 | Prorated | $0.00 | $0.00 | $0.00 | $0.00 | $0.00 |
| **Entry gates** | **$0.00** | **Prorated** | **$23,125.00** | **$23,125.00** | **$0.00** | **$7,135.00** | **$15,990.00** |
| Select gates | $0.00 | Prorated | $6,400.00 | $6,400.00 | $0.00 | $6,400.00 | $0.00 |
| Purchase gates | $0.00 | Start | $13,470.00 | $13,470.00 | $0.00 | $735.00 | $12,735.00 |
| Install gates | $0.00 | Prorated | $2,400.00 | $2,400.00 | $0.00 | $0.00 | $2,400.00 |
| Install card readers | $0.00 | Prorated | $427.50 | $427.50 | $0.00 | $0.00 | $427.50 |
| Install infrared photo c | $0.00 | Prorated | $427.50 | $427.50 | $0.00 | $0.00 | $427.50 |
| Perimeter Complete | $0.00 | Prorated | $0.00 | $0.00 | $0.00 | $0.00 | $0.00 |



**FIGURE 16-6:** Cost table.

WARNING

Project assumes that the remaining costs will stay at the budgeted amount. In other words, from the total budgeted amount for the work, it assumes that the remaining work will be completed on budget, and it adds the actual costs to date to produce the total cost. Rarely do future estimates end on budget when the work to date is in a cost-overrun situation, so don't rely on the information in the Total Cost column.

In Figure 16-7, the Variance table shows whether the schedule is on track or has variances (much like the Cost table shows cost variances). The table shows the variance between the start and finish of tasks and the baseline start and finish dates. Notice in the example that the Dig Trench and Install Cinder Blocks tasks finished early, which allowed the Select Gates task to start early as well.

| Task Name | Start | Finish | Baseline Start | Baseline Finish | Start Var. | Finish Var. |
|---|---|---|---|---|---|---|
| **Perimeter** | **Mon 1/3/22** | **Fri 4/8/22** | **Mon 1/3/22** | **Fri 4/8/22** | **0 days** | **0 days** |
| **Walls** | **Mon 1/3/22** | **Fri 3/11/22** | **Mon 1/3/22** | **Fri 3/11/22** | **0 days** | **0 days** |
| Dig trench | Mon 1/3/22 | Fri 1/14/22 | Mon 1/3/22 | Mon 1/17/22 | 0 days | -0.88 days |
| Install footing | Mon 1/17/22 | Tue 1/25/22 | Mon 1/17/22 | Tue 1/25/22 | -0.88 days | 0 days |
| Install cinder blocks | Mon 1/24/22 | Mon 2/14/22 | Mon 1/24/22 | Wed 2/16/22 | 0 days | -2.13 days |
| Apply stucco | Mon 2/14/22 | Mon 2/21/22 | Mon 2/14/22 | Mon 2/21/22 | 0 days | 0 days |
| Paint | Mon 2/21/22 | Mon 2/28/22 | Tue 2/22/22 | Mon 2/28/22 | -0.13 days | 0 days |
| Apply finishes | Mon 2/28/22 | Fri 3/4/22 | Mon 2/28/22 | Fri 3/4/22 | 0 days | 0 days |
| Wall contingency | Mon 3/7/22 | Fri 3/11/22 | Mon 3/7/22 | Fri 3/11/22 | 0 days | 0 days |
| Walls complete | Fri 3/11/22 | Fri 3/11/22 | Fri 3/11/22 | Fri 3/11/22 | 0 days | 0 days |
| **Entry gates** | **Wed 2/16/22** | **Fri 4/8/22** | **Mon 2/14/22** | **Fri 4/8/22** | **1.13 days** | **0 days** |
| Select gates | Wed 2/16/22 | Fri 2/25/22 | Mon 2/14/22 | Mon 2/28/22 | 1.13 days | -0.88 days |
| Purchase gates | Mon 2/28/22 | Mon 3/21/22 | Mon 2/28/22 | Mon 3/21/22 | 0 days | 0 days |
| Install gates | Wed 3/30/22 | Fri 4/1/22 | Wed 3/30/22 | Fri 4/1/22 | 0 days | 0 days |
| Install card readers | Mon 4/4/22 | Mon 4/4/22 | Mon 4/4/22 | Mon 4/4/22 | 0 days | 0 days |
| Install infrared photo devices | Fri 4/8/22 | Fri 4/8/22 | Fri 4/8/22 | Fri 4/8/22 | 0 days | 0 days |
| Perimeter Complete | Fri 4/8/22 | Fri 4/8/22 | Fri 4/8/22 | Fri 4/8/22 | 0 days | 0 days |

**FIGURE 16-7:** Variance table.

REMEMBER

The critical path is the series of tasks that must be completed on time for the project to meet its final finish date.

REMEMBER

If you created contingency reserve to help handle unexpected delays, the total variance shown in the Variance table tells you how many days you may have to deduct from the contingency duration to get back on track. You can read more in Chapter 17 about making adjustments for delays and cost overruns and about contingency reserve in Chapter 12.

# Tracking Progress Using Earned Value Management

On large projects, especially large government projects, project managers are required to track progress using the earned value management (EVM) method. I won't bore you with a detailed discussion of EVM, but you should become familiar with some of the most common terms and calculations surrounding this concept because many organizations require information on these specific numbers in project reports. Some common terms are defined in this list:

- **Planned value (PV):** The budgeted (or baseline) cost of tasks based on the cost of resources assigned to the task, plus any fixed costs associated with the tasks, at the point of measurement. In some circles, planned value is also known as the budgeted cost of work scheduled, or BCWS. For example, if you're installing 100 fixtures, the baseline cost to install a fixture is $100, and you estimate that you can install 5 fixtures per day, the cumulative planned value for Day 3 is $1,500 ($500 each day for Day 1, Day 2, and Day 3).
- **Earned value (EV):** The value of the work you've completed, expressed in dollars. For example, if you've installed only 13 fixtures at the end of Day 3, the earned value is $1,300 (each installation was valued at $100, and 13 × $100 = $1,300). In some circles, earned value is referred to as the budgeted cost of work performed, or BCWP.
- **Actual cost (AC):** A calculation that includes tracked resource hours or units expended on the task plus fixed costs. In the example, assume that you had to pay overtime to complete some of the work and that the actual costs paid to the installers at the end of Day 3 are $1,650. AC is also known as actual cost of work performed, or ACWP.
- **Budget at completion (BAC):** The sum of the planned value. In the example, the BAC of the project is 100 fixtures × $100, or $10,000.
- **Estimate at completion (EAC):** A forecast of the total cost of the project. Project calculates that EAC = BAC – EV + AC. EVM practitioners cringe at the assumption that future costs will stay on budget. However, Project isn't an EVM tool. If you need to get serious about using EVM, you can buy a plug-in that accurately calculates multiple EACs based on assumptions of future performance.
- **Variance at completion (VAC):** The difference between the budget at completion (BAC) and the estimate at completion (EAC) is VAC = BAC – EAC.
- **Cost variance (CV):** The difference between earned value and actual costs. The equation is CV = EV – AC. This number is expressed as a negative number if the project is over budget and as a positive number if you're under budget.
- **Schedule variance (SV):** The difference between earned value and planned value. The equation is SV = EV – PV. This number is expressed as a negative number if you're accomplishing less work than planned; it's a positive number if you're accomplishing more work than planned. Be aware of saying aloud that you're behind or ahead of schedule. If the work isn't on the critical path, the schedule performance doesn't necessarily indicate an ahead-of-schedule status or a behind-schedule status.

## Viewing the Earned Value table

To see the Earned Value table, as shown in Figure 16-8, select the View tab on the Ribbon, click Tables in the Data group, select More Tables, and choose Earned Value. As you can see, there is a lot of data there! You can also look at tables that focus only on Earned Value Cost Indicators or Earned Value Schedule Indicators.

| Task Name | Planned Value - PV (BCWS) | Earned Value - EV (BCWP) | AC (ACWP) | SV | CV | EAC | BAC | VAC |
|---|---|---|---|---|---|---|---|---|
| Perimeter | $222,425.00 | $232,438.92 | $215,345.00 | ($1,986.08) | $17,093.92 | $235,211.16 | $240,775.00 | $5,563.84 |
| Walls | $216,025.00 | $214,666.67 | $210,225.00 | ($1,358.33) | $4,441.67 | $213,146.61 | $217,650.00 | $4,503.39 |
| Dig trench | $3,000.00 | $3,000.00 | $3,050.00 | $0.00 | ($50.00) | $3,050.00 | $3,000.00 | ($50.00) |
| Install footing | $2,700.00 | $2,366.67 | $2,700.00 | ($333.33) | ($333.33) | $3,080.28 | $2,700.00 | ($380.28) |
| Install cinder blocks | $158,200.00 | $158,200.00 | $151,000.00 | $0.00 | $7,200.00 | $151,000.00 | $158,200.00 | $7,200.00 |
| Apply stucco | $50,125.00 | $50,125.00 | $52,500.00 | $0.00 | ($2,375.00) | $52,500.00 | $50,125.00 | ($2,375.00) |
| Paint | $2,000.00 | $975.00 | $975.00 | ($1,025.00) | $0.00 | $2,000.00 | $2,000.00 | $0.00 |
| Apply finishes | $0.00 | $0.00 | $0.00 | $0.00 | $0.00 | $1,625.00 | $1,625.00 | $0.00 |
| Wall contingency | $0.00 | $0.00 | $0.00 | $0.00 | $0.00 | $0.00 | $0.00 | $0.00 |
| Walls complete | $0.00 | $0.00 | $0.00 | $0.00 | $0.00 | $0.00 | $0.00 | $0.00 |
| Entry gates | $6,400.00 | $17,772.25 | $5,120.00 | ($627.75) | $12,652.25 | $20,511.93 | $23,125.00 | $2,613.07 |
| Select gates | $6,400.00 | $5,760.00 | $5,120.00 | ($640.00) | $640.00 | $5,688.89 | $6,400.00 | $711.11 |
| Purchase gates | $0.00 | $12,012.25 | $0.00 | $12.25 | $12,012.25 | $13,470.00 | $13,470.00 | $0.00 |
| Install gates | $0.00 | $0.00 | $0.00 | $0.00 | $0.00 | $2,400.00 | $2,400.00 | $0.00 |
| Install card readers | $0.00 | $0.00 | $0.00 | $0.00 | $0.00 | $427.50 | $427.50 | $0.00 |
| Install infrared photo devices | $0.00 | $0.00 | $0.00 | $0.00 | $0.00 | $427.50 | $427.50 | $0.00 |
| Perimeter Complete | $0.00 | $0.00 | $0.00 | $0.00 | $0.00 | $0.00 | $0.00 | $0.00 |
| Equipment | $95.00 | $95.00 | $95.00 | $0.00 | $0.00 | $63,500.00 | $63,500.00 | $0.00 |
| Asset Management | $0.00 | $0.00 | $0.00 | $0.00 | $0.00 | $0.00 | $0.00 | $0.00 |
| Operations Readiness | $0.00 | $0.00 | $0.00 | $0.00 | $0.00 | $20,800.00 | $20,800.00 | $0.00 |

**FIGURE 16-8:** Earned Value table.



## Earned value options

As I mention earlier in this chapter, usually only large projects use earned value management. To find the settings to customize how earned value works in the project, select the File tab, click the Options button, and select Advanced. Scroll down until you see Earned Value Options for This Project. Figure 16-9 shows you the two sample settings for calculating earned value.

The Default Task Earned Value Method setting provides these two choices:

» **% Complete:** This setting calculates earned value using the percent complete that you record on each task. The setting assumes that, on a task that's halfway complete, half the work hours have been used.

- **Physical % Complete:** Use this setting if you want to manually enter a percentage of completion not based on a straight percent-complete calculation. For example, if you have a four-week task to complete a mail survey, 50 percent of the effort may happen in the first 25 percent of the duration of the project: Design, print, and mail the survey. Nothing happens for two weeks while you wait for responses and then you see a flurry of activity when the responses are returned to you. So a straight calculation that 50 percent of the task is completed 50 percent of the way through isn't accurate. If your projects have a lot of tasks of this type, you may consider changing the settings to use this method. Then you can display the Physical % Complete column in the Gantt Chart sheet and enter more accurate (in your opinion) percent-complete information for each task.

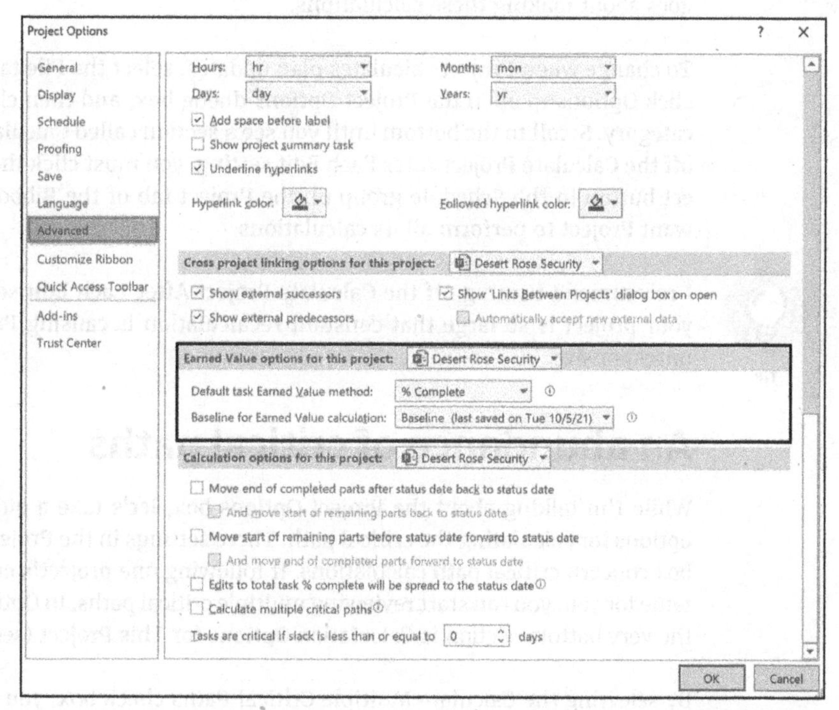


**FIGURE 16-9:** Two settings to calculate earned value.

© John Wiley & Sons, Inc.

The second setting in the Earned Value Options for This Project section of the Project Options dialog box is the Baseline for Earned Value Calculation drop-down list. The baseline you use to calculate earned value is essentially setting the planned value for the project. If you're using multiple baselines — such as one for reporting and one for managing — pay close attention to the baseline against which you choose to measure the earned value. Choose any of the 11 possible baselines you may have saved in the project. After you make these two choices, click OK to close the Project Options dialog box.

# Calculating behind the Scenes

While you're happily entering resource hours and fixed costs into the project, Project is busy making calculations that can shift around task timing and resource workload in the plan. These calculations relate to how tasks are updated, how the critical path is determined, and how earned value is calculated. If you're a control freak, you'll be happy to know that you can, to some extent, control how Project goes about making these calculations.

To change when Project calculates plan updates, select the File tab on the Ribbon, click Options to open the Project Options dialog box, and then click the Schedule category. Scroll to the bottom until you see a section called Calculation. If you turn off the Calculate Project After Each Edit setting, you must click the Calculate Project button in the Schedule group on the Project tab of the Ribbon whenever you want Project to perform all its calculations.

TIP

I recommend turning off the Calculate Project After Each Edit setting only when your project is so large that constant recalculation is causing Project to become unresponsive.

## An abundance of critical paths

While I'm talking about the Project Options box, let's take a quick look at some options for calculating the critical path. Three settings in the Project Options dialog box concern critical path calculations. If following one project's critical path is too tame for you, you can start reviewing multiple critical paths. In Options ⇨ Advanced, the very bottom section is Calculation Options for This Project (see Figure 16-10).

By selecting the Calculate Multiple Critical Paths check box, you set up Project to calculate a different critical path for each set of tasks in the project. Doing so can be helpful if you want to identify tasks that, if delayed, will cause you to miss the final project deadline or the goals of a single phase in the project.

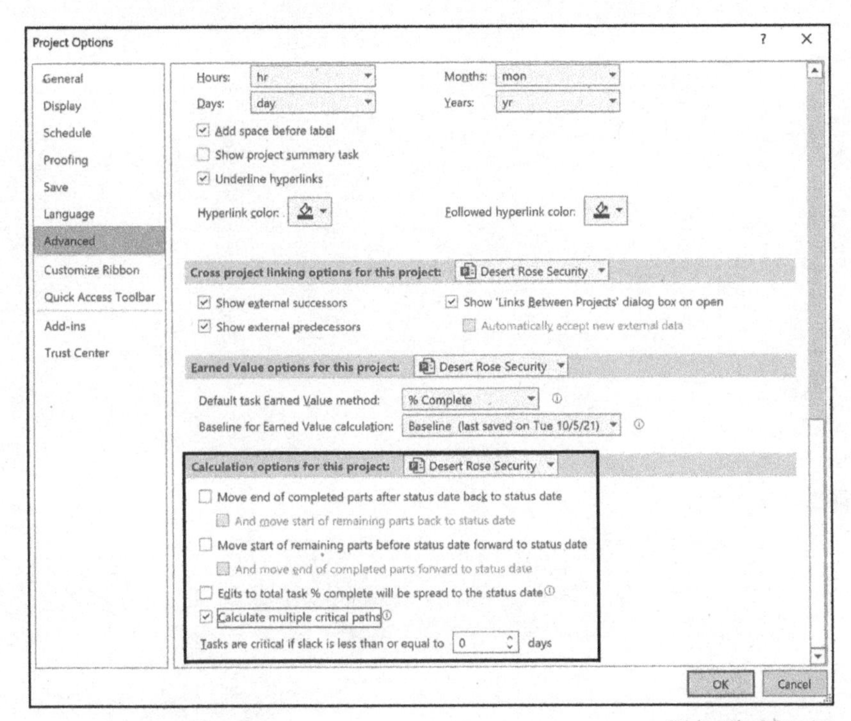


**FIGURE 16-10:** Selecting multiple critical paths.

*© John Wiley & Sons, Inc.*

You can establish which factors add a task to the critical path, by specifying the number of days of slack the critical tasks may have. Tasks with no slack are, by default, on the critical path. However, you can change this situation if you want to be alerted that tasks with only one day of slack are critical — figuring that one day isn't much padding and that these tasks are still in jeopardy. The setting labeled Tasks Are Critical If Slack Is Less Than or Equal to (x) Days is immediately beneath the Calculate Multiple Critical Paths check box in the Advanced category.

Finally, in Options ⇨ Schedule, you'll find the Inserted Projects Are Calculated Like Summary Tasks check box in the Calculation Options for This Project section at the bottom of the category. This setting is straightforward: If you insert another project as a task in the project, you allow Project to calculate one critical path for the entire project. If you don't select it, any projects that you insert are treated as outsiders — that is, they aren't considered in the master project's critical-path calculations. If an inserted project won't affect the project's timing, you may deselect this option.

IN THIS CHAPTER

» Analyzing plan versions and notes to understand what went wrong

» Experimenting with what-if scenarios

» Benefitting from more time or more people

» Adjusting the timing going forward

# Chapter 17

# You're Behind — Now What?

Let's face it: A project rarely proceeds as planned. Every project has issues (situations that must be handled before the project can move forward) and risks (uncertain events that can prevent a project objective from being met). One day the project is doing just fine, and the next day it's $75,000 over budget. Or it appears that you'll miss the drop-dead finish date by two weeks.

The first step is to find the cause and then fix it so that you can move forward and try to save the project. To try to save the day when the project rolls off the track, you need to analyze the options and make tough choices. Project can help you analyze possible solutions and anticipate the likely results. After you decide what to do, you can update Project to reflect the solutions.

## Using Project with Risk and Issue Logs

You can proactively document problems on the project in a number of ways, especially if you create and maintain an issue log and a risk register on the project. In Chapter 12, I describe how to handle risk, assign contingency reserve, and update the project schedule to include the work, resources, time, and funding necessary to implement risk responses.

## Documenting issues

An *issue* is a situation that you have to deal with before you can move forward, such as a risk that develops, a decision that has to be made, or an uncooperative stakeholder. Whatever the case, you should record issues in an *issue log* so that you can assign accountability for managing and resolving them in a timely manner.

You can also use Project to run what-if scenarios and develop alternative approaches for responding to issues or risks. You may have developed multiple baselines (described in Chapter 14) to account for various outcomes. It is a good practice to record task notes for those tasks affected by a risk or an issue. (Information on task notes is in Chapter 3.) Storing interim plans, multiple baselines, and task notes in Project helps you more easily explain messy situations to the powers that be.

## Printing interim plans and baselines

You can use interim plans and multiple baselines to demonstrate how you've made adjustments for risks or issues. Using these two items indicates that you were attuned to potential problems. You should also update the boss and other stakeholders by generating printouts or reports reflecting potential major issues and risks. I talk more about reports in Chapter 18.

REMEMBER

A *baseline* saves pertinent project data, including start and finish dates, duration, work, and cost; an *interim plan* saves only the start and finish dates of tasks in the project. Chapter 14 describes interim plans and baselines in depth.

To view multiple baselines, follow these steps:

1. **Click the Task tab. In the View section, click the down arrow for a list of Views.**
2. **At the bottom of the list, choose More Views.**
3. **In the More Views dialog box, select Multiple Baselines Gantt.**
4. **Click Apply.**

Multiple Baselines Gantt view shows the original Baseline, Baseline 1, and Baseline 2. To customize what you see by including baselines and interim plans in whichever way you want to see them, follow these steps:

1. **Display Task Sheet view.**
2. **Scroll the sheet pane to the right and click the Add New Column column heading.**

   The menu listing available fields appears, as shown in Figure 17-1.

3. **Scroll down and click the field you want to insert.**

   For example, you may choose one of 14 specific Baseline fields for Baseline through Baseline 10. Figure 17-1 shows the available fields for Baseline and Baseline 1. If you want to look at interim plans, you can choose Start 1–10 and Finish 1–10.

4. **If necessary, repeat Steps 2 and 3 to display additional columns.**

Baseline Budget Cost
Baseline Budget Work
Baseline Cost
Baseline Deliverable Finish
Baseline Deliverable Start
Baseline Duration
Baseline Estimated Duration
Baseline Estimated Finish
Baseline Estimated Start
Baseline Finish
Baseline Fixed Cost
Baseline Fixed Cost Accrual
Baseline Start
Baseline Work
Baseline1 Budget Cost
Baseline1 Budget Work
Baseline1 Cost
Baseline1 Deliverable Finish
Baseline1 Deliverable Start
Baseline1 Duration
Baseline1 Estimated Duration
Baseline1 Estimated Finish
Baseline1 Estimated Start
Baseline1 Finish
Baseline1 Fixed Cost
Baseline1 Fixed Cost Accrual
Baseline1 Start
Baseline1 Work

**FIGURE 17-1:** Adding more columns.



# What-If Scenarios

Sometimes you need a fresh perspective to help you see and solve a problem. Using the filtering and sorting features in Project, you can slice and dice various aspects of the project somewhat differently to get that fresh perspective.

You can also use tools, such as resource leveling, to solve resource conflicts. Resource leveling may not always solve problems to your satisfaction, but it's a good way to let Project show you a what-if scenario that may solve resource problems instantly.

**REMEMBER**

See Chapter 12 for more information about how to use resource leveling.

## Sorting tasks

Sometimes when the project isn't going as planned, it's time to sort tasks. In Project, you can sort tasks by several criteria, including start date, finish date, priority, and cost.

To determine how sorting can help you, consider these examples:

» **To cut costs:** When you sort tasks by cost, you can focus on the most expensive tasks first, to see whether you have room to trim useful-but-pricey items.

» **To review task timing:** When you sort by duration in descending order, you can see the longest tasks first.

To apply a preset sorting order, simply choose View⇨Data⇨Sort. The Sort button has an A over a Z with a down arrow next to it. From the Sort drop-down menu, you can choose an option such as By Start Date or By Cost.

If you want to see additional sort criteria or to sort by more than one criterion, follow these steps:

1. **Choose View⇨Data⇨Sort⇨Sort By.**

   The Sort dialog box appears, as shown in Figure 17-2. You can use it to sort by as many as three criteria in ascending or descending order. The first field shows the ID field by default. The remaining Then By fields have no values selected.

2. **In the Sort By list, select a criterion to sort by.**
3. **Select either Ascending (to sort from lowest to highest) or Descending (to sort from highest to lowest).**

   In a date field, the sorting order is from soonest to latest and from latest to soonest, respectively; in a text field, alphabetical is the order.

4. **(Optional) If you want a second criterion, open the first Then By drop-down list and make a selection.**

   For example, if you choose to sort first by cost and then by type, tasks are sorted from least expensive to most expensive and then (within each cost level) by type (Fixed Duration, Fixed Units, and Fixed Work).

5. **(Optional) If you want to add a third criterion, open the second Then By drop-down list and make a selection.**
6. **Click the Sort button.**

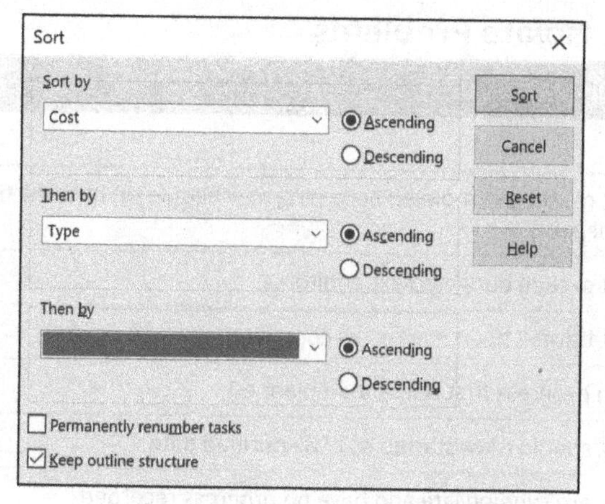


**FIGURE 17-2:** Choosing criteria for sorting.



To return to the original task order, choose View ⇨ Data ⇨ Sort ⇨ By ID. Tasks are now back in *task ID number* order, which is the standard sorting criterion in Project.

## Filtering

Filtering, as described in Chapter 11, is a good way to focus on the areas of the project that are at risk. Especially in larger projects, where it can be difficult to scan hundreds of tasks and notice which ones are running late or over budget, filters can home in on exactly where the trouble lies. Filters are on the View tab, in the Data group.

You can simply highlight tasks that meet the criteria onscreen by working with the Highlight drop-down menu instead of the Filtering drop-down menu.

Table 17-1 lists filters that are useful when you're trying to identify and solve problems with the schedule. You won't see all these choices when you click the Filter button. You need to select the More Filters option near the bottom of the list. Then select the Task radio button to see task-related filters or the Resource radio button to see resource-related filters.

You can export Project information to a program such as Excel so that you can use analysis tools, such as pivot tables. If this idea excites you, check out *Excel For Dummies* by Greg Harvey (Wiley) or *Excel Data Analysis For Dummies* by Paul McFedries (Wiley).

**TABLE 17-1** Filters That Isolate Problems

| Filter Name | What It Displays |
|---|---|
| **Task Filters** | |
| Critical | Tasks that must be completed according to schedule to meet the final deadline (the critical path) |
| Cost Overbudget | Tasks that exceed budgeted expenditures |
| Incomplete Tasks | Tasks that haven't been marked as complete |
| Late Tasks | Tasks with progress that is less than planned |
| Should Start By | Tasks that should have started as of a specified date |
| Slipped/Late Progress | Tasks that are running late and have no progress recorded |
| **Resource Filters** | |
| Overallocated Resources | Tasks with resources assigned that are overbooked at some point during the life of the task |
| Slipping Assignments | Tasks involving resource work that should have already begun |
| Work Overbudget | The number of work hours that have exceeded the original estimate |

## Examining the critical path

One useful filter in Project is named Critical; it displays or highlights all tasks that are on the critical path. If you're running late, knowing which tasks can't slip helps you identify areas that have no room for delay — and, conversely, areas where you can delay noncritical tasks and still meet the deadline. You may use the Critical filter to help determine how to free overallocated resources or get a task that's running late back on track.

REMEMBER

Ensure that the Gantt chart shows the Gantt bars for critical tasks in red. To do so, make sure you're in the Gantt Chart view; then select the Gantt Chart Format tab on the Ribbon and ensure that the Critical Tasks check box is selected in the Bar Styles group.

You can look at the critical path in any Gantt Chart or Network Diagram view. Figure 17-3 shows Gantt Chart view of a project sorted for the critical path. Figure 17-4 shows Network Diagram view (with the boxes collapsed) with the same filter applied.

TIP

If you need a closer look at task timing, modify the timescale display to use smaller increments, such as days or hours. To do so, right-click the timescale itself and then choose Timescale. You can also click View and then, in the Zoom group, click the Timescale and select your preferred timescale. Or you can click the Zoom drop-down arrow and select Zoom In. You can also use the slider at the bottom-right side of the screen. It's less precise, but it's a quick way to zoom to the level you need.

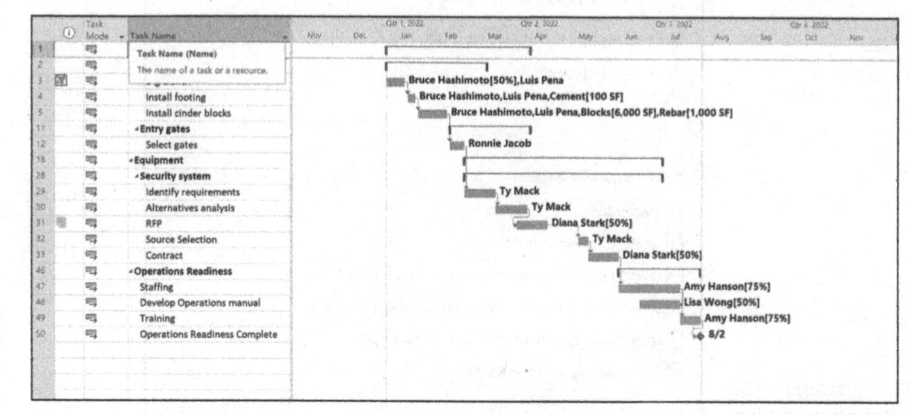


**FIGURE 17-3:** Gantt Chart view filtered for the critical path.

© John Wiley & Sons, Inc.

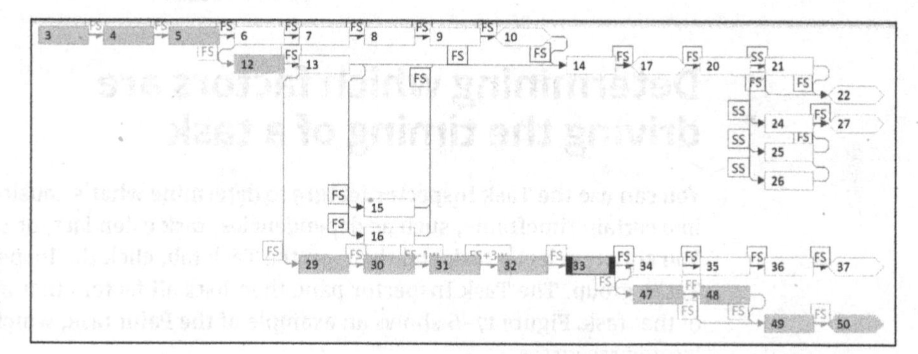

**FIGURE 17-4:** Network Diagram view of the critical path.

© John Wiley & Sons, Inc.

# Using resource leveling (again)

If you use manual resource leveling early in the project to solve resource conflicts prior to baselining, try it again to help you when you need to update the schedule. With changes to tasks and tracked activity, resource leveling may uncover new options to solve conflicts.

If resource leveling is set to Automatic, Project automatically performs this calculation every time you modify the schedule. To see whether the option is set to Automatic or Manual, choose Resource ⇨ Level ⇨ Leveling Options to open the Resource Leveling dialog box, as shown in Figure 17-5. If the Manual radio button is selected, close the dialog box, then click Level All in the Level group to run resource leveling.

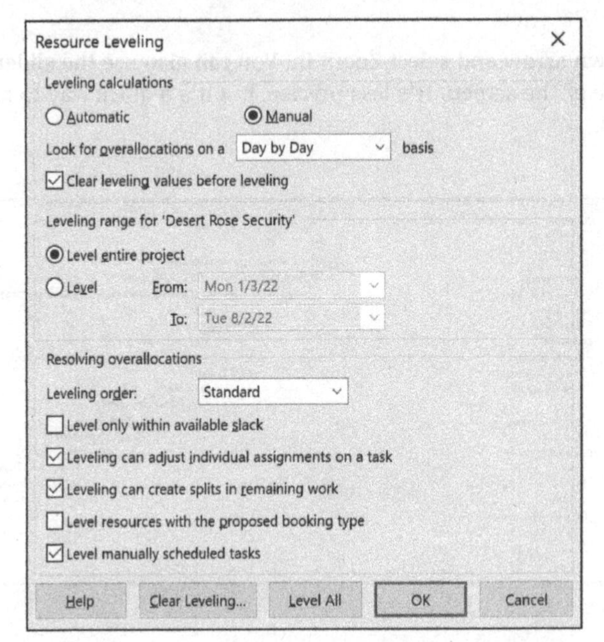


**FIGURE 17-5:** Resource Leveling dialog box.

*© John Wiley & Sons, Inc.*

## Determining which factors are driving the timing of a task

You can use the Task Inspector feature to determine what's causing tasks to occur in a certain timeframe, such as dependencies, task calendars, or task constraints. You simply select a task and then, on the Task tab, click the Inspect button in the Tasks group. The Task Inspector pane then lists all factors that affect the timing of that task. Figure 17-6 shows an example of the Paint task, which shows overallocated resources.

By using this feature (read more about it in Chapter 11), you can determine whether a task that you want to take place earlier can do so if you remove a dependency or constraint that's affecting it. For example, when developing the schedule you may have arranged a series of tasks as finish-to-start (the successor task starts after the predecessor finishes), but now you want to accelerate the work. You may want to consider bringing in an extra resource for the Walls work package, for example. This will allow you to fast track some of the work. As a reminder, fast tracking is performing some or all of the work in parallel rather than sequentially (i.e., starting the successor task before the predecessor is done). By understanding what's driving the timing of the task, you can better search for a solution if that timing is causing problems.

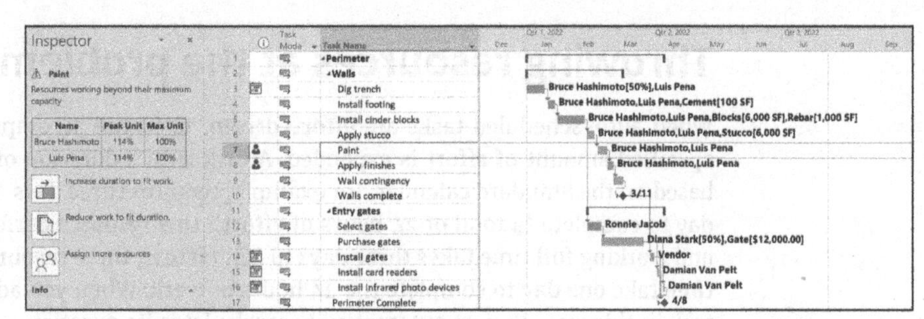

**FIGURE 17-6:** Task Inspector pane.

© John Wiley & Sons, Inc.

# How Adding People or Time Affects the Project

**Human nature in a corporate environment often causes people to want to throw money and people at projects. In certain cases, that instinct is on target; however, you aren't always able to draw on an endless supply of resources or time. For this reason, you may have to experiment with a combination of options that involve time and resources.**

## Hurrying up and making modifications

Saving time in Project requires that a task be completed earlier than scheduled. You're likely to find that making this adjustment is similar to trying to solve an intricate puzzle: Correct one piece, and another one begins causing aggravation.

To accomplish work faster, you have these two options:

- **Find more resources to help with tasks that are auto-scheduled.** Adding people often requires money, so even if the schedule gets back on track, it will cost you.
- **Modify the scope of tasks.** This strategy may have an effect on the task quality. If you make two inspections rather than three or you shorten the QA cycle by a week, you may run the risk of causing problems down the road.

Changing the timing of tasks and shifting dependencies uses up slack to make up for delays, but it may leave you with no wiggle room. The next time a problem crops up, you'll be up against the wall with no slack.

In reality, combining small modifications for both time and money can often help save the day.

## Throwing resources at the problem

When auto-scheduled tasks are effort-driven, tasks are accomplished when the specified amount of effort is expended. A task with a duration of three days and based on the Standard calendar, for example, requires three days × eight hours per day to complete (a total of 24 hours of effort). One resource performing this task and working full time takes three days to complete it; three resources working full time take one day to complete the 24 hours of work. When you add resources to a task in this way, Project automatically recalculates its duration.

REMEMBER

If you want Project to reduce the duration of a task when you assign more work resources to it, the task must be auto-scheduled, and effort-driven scheduling must be turned on.

## Changing how resources are assigned

Beyond simply adding resources to a task, you can modify existing assignments. On any given project, you may have dozens (or even hundreds) of resources working on tasks. All these people are working according to their working calendars, the percentage of resource time assigned to particular tasks, and their ability to do the job. Take a look at how folks were assigned to begin with to see whether you can save time or money by modifying those assignments.

You can modify assignments in several ways:

- If someone is working at, say, only 50 percent capacity on a task, explore whether you can increase their working time, at least for the time they're attending to tasks on the critical path.
- If a resource is available who can perform a given task more quickly, switch resources on the task and shorten its duration. Remember that the Team Planner view can be helpful when making this type of switch. A more skilled person may cost more, so take the time/cost trade-off into consideration.
- Have certain resources work overtime or let them be overbooked at various points during the project. You may have modified an overbooked resource's assignments earlier to eliminate a conflict, but now you find that you have no choice other than to have the resource work an occasional 10- or 12-hour day.

## Calculating the consequences of schedule modifications

Before you get carried away from making changes to resources, stop and think: Adding resources to effort-driven, auto-scheduled tasks can shrink the task and help the project get back on track. However, depending on the resources' hourly rates, this approach may cost more.

Assigning three people to work on a task doesn't necessarily shrink the duration of the task geometrically. That's because those people have to coordinate their efforts, hold meetings, and spend more time communicating than they would if they were working on the task alone. If you add resources, Project shrinks the task geometrically: Consider adding a little time to the task to accommodate the inefficiencies of multiple resources.

Another concern about adding resources to tasks is that it may cause more resource conflicts, with people who are already busy being overbooked on too many tasks that happen in the same timeframe. If you have the resources, and they have the skills and the time, though, beefing up the workforce is definitely one way to perform certain tasks more quickly.

REMEMBER

To add resources to a task, you can use the Resources tab in the Task Information dialog box, or choose the task and click the Assign Resources button in the Assignments group on the Resource tab of the Ribbon.

## Shifting dependencies and task timing

Time is a project manager's greatest enemy. There's never enough time, and the time you have is eaten up quickly.

Follow these suggestions to modify task timing and save time:

- **Delete a task.** You heard me. If a task represents a step that can be skipped and no actual work for it is tracked, simply delete the task. It doesn't happen often, but sometimes — after you rethink the project — you realize that a few actions are unnecessary or have already been handled by someone else. If you prefer, you also can mark the task as inactive rather than deleting it.
- **Adjust dependencies.** Perhaps the revision of the operations manual can start a few days before *all* the feedback is returned. Or perhaps installing the gates and card readers can take place at the same time rather than one after the other (assuming that the resources can stay out of each other's way). Use the Predecessors tab in the Task Information dialog box, shown in Figure 17-7, to modify dependencies. In the Lag column, you can enter a negative number to allow tasks to overlap.
- **Modify constraints.** Perhaps you've set a task to start no earlier than the first of the year because you don't want to spend money on it until the new fiscal year budget kicks in. To save time, consider whether you can allow it to start a week before the end of the year and bill the costs in January. Examine any constraint of this type — especially those created to verify the timing logic.

» **Verify external dependencies.** If you've created an external link to a task in another project and set dependencies with tasks in the project, see whether the other project manager can speed up certain tasks. Or if the timing relationship isn't absolutely critical, delete the link to the other project. It may be slowing you down more than you realize.

» **Change the scheduling method or move the task.** Remember that Project usually cannot automatically reschedule a task based on dependency changes if the task is manually scheduled. You can either change the task to be auto-scheduled by using the Auto Schedule button in the Tasks group on the Task tab of the Ribbon, or you can select a manually scheduled task and then click the Respect Links button in the Schedule group on the Task tab to move it to its proper timeframe. Or you can use the Move drop-down list on the Tasks group on the Task tab to reschedule a task to a particular earlier or future time slot, or to one when resources are available.

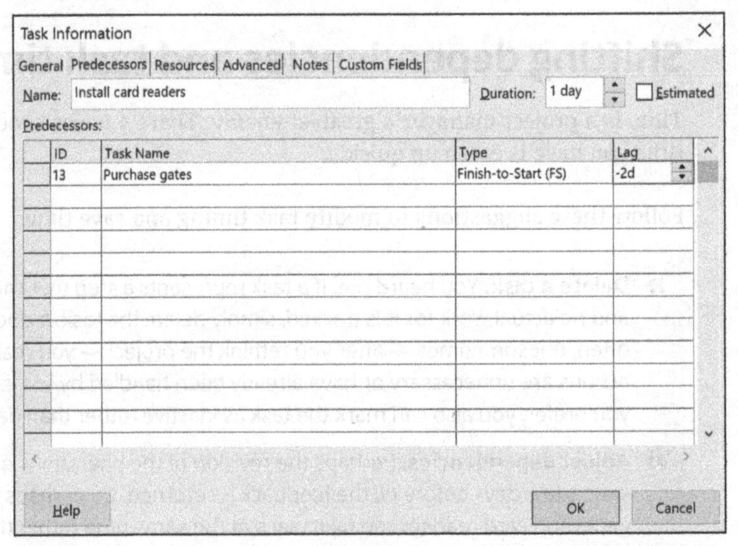


**FIGURE 17-7:** Modifying task dependencies.

© *John Wiley & Sons, Inc.*

**If you've set resource leveling to Automatic, Project may have delayed certain tasks until overbooked resources are freed up. On the Resource tab, click the Leveling Options button in the Level group and change the setting to Manual.**

# When All Else Fails

Okay. You've monkeyed with resource assignments and shifted around task dependencies to save time, and you've deleted tasks and assigned less expensive resources to save money. Still, it's not enough. In this scenario, you have to say to the boss, "You can have it on time, you can have it on budget, or you can have quality work. Choose two."

REMEMBER

If the boss throws money at you, by all means add resources to tasks, as discussed in the earlier section "Throwing resources at the problem." But keep in mind that not all tasks are effort driven. Some tasks aren't shortened by adding resources.

## Taking the time you need

If the boss is willing to give you more time, grab it. However, when you do, you have to update the project a few ways:

- **Modify task durations.** Add time to tasks that are running late. In Project, you do this by increasing their durations or moving out their start or end date.
- **Review task constraints.** If you've specified that certain tasks can finish no later than a certain date, but now you're extending the project finish date three months, you may be able to remove or adjust those original constraints accordingly.
- **Adjust the contingency reserve.** If you have a contingency reserve task, you can simply add to its duration, giving more waffle room to all other tasks. (See Chapter 12 for more about contingency reserve.) If you're using reserve to stay on schedule, you can reduce the amount of contingency reserve by the amount you're using.

After you adjust the time, you should make sure that the new timing of tasks doesn't cause new resource conflicts by looking at Resource Graph view, and then reset the baseline to reflect the new schedule. To reset a baseline, select Project ⇨ Schedule ⇨ Set Baseline ⇨ Set Baseline. Choose Baseline 1–10 in the Set Baseline dialog box to save to a different baseline and preserve the original. Click OK after choosing a baseline.

TIP

Inform team members of the new timing and provide them an updated version of the plan.

## Finding ways to cut corners

If your manager tells you to cut some corners that sacrifice quality, you have license to modify the quality of the deliverables. You can eliminate tasks that may ensure higher quality, such as a final proofreading of the operations manual. You can hire less experienced and less expensive workers or you can use cheaper material.

To cut corners in Project, you have to:

- **Take fewer steps:** Delete tasks. Click the ID of the task in Gantt Chart view and click Delete. Alternatively, you can mark the task as Inactive. Don't forget to review any dependencies that were broken because you deleted tasks.
- **Use less expensive team resources:** Delete one set of resource assignments and assign other resources to tasks in the Assign Resources dialog box.
- **Use less expensive materials:** Change the unit price of material resources you've created in Resource Sheet view. Or shop around for better pricing on items that are set up as cost resources, and enter the lower prices of tasks to which you've assigned those resources by using the Cost column in the Assign Resources dialog box, as shown in Figure 17-8. To get to the Assign Resources dialog box, right-click a task or a resource. From the drop-down menu, choose Assign Resources.

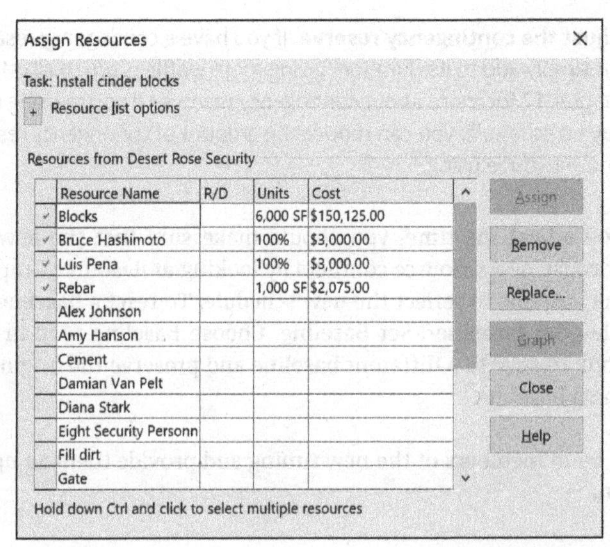


**FIGURE 17-8:** Modifying a resource cost.

You can also take a more sweeping approach: Simply redefine the goal of the project. If the goal were to launch a new product line, perhaps you can modify the goal to simply manage the design of the new product and then leave the launch to a later date or another project manager. If you were supposed to produce 10,000 widgets, see whether the company can make do with 7,500. To make this type of change, you may have to slice and dice entire phases of the project — or even start from scratch and build a new plan.

TIP

Save the revised project schedule with a new name to provide a head start on the revised project schedule. Clear the baseline (choose Project⇨Schedule⇨Set Baseline⇨Clear Baseline, and then click OK), make modifications, and then save a new baseline.

IN THIS CHAPTER

» Generating standard reports

» Creating custom reports

» Dazzling readers with visual reports

» Adding graphics and formatting to reports

» Optimizing printer settings

» Sharing a snapshot of the Timeline

» Enhancing reports with copy and paste

# Chapter 18
# Spreading the News: Reporting

Here it is — the big payoff. It's your reward for entering all those task names and resource hourly rates, and for tracking activity on dozens of tasks during those late-hour sessions in the first hectic weeks of the project. You're finally ready to print a report or another type of project data, receiving tangible benefits from Project that you can hand out at meetings and show off to impress the boss.

Reports help you communicate information about projects by conveying schedule progress, resource assignments, cost accumulation, and activities that are in progress or scheduled soon. You can take advantage of built-in reports or create new reports to include the data that's most relevant to you. Dashboard reports and visual reports in Project offer graphical possibilities to help paint a picture of the progress.

Knowing that you want to impress people, Project also makes it possible to apply certain themes, images, shapes, charts, and tables to reports to help make the point.

A report isn't the only tool you use to communicate with. You can customize headers, footers, and legends, and insert graphics into the schedule to make a presentation memorable. Other capabilities, such as customizing and copying Timeline view, enable you to incorporate Project data into reports in other programs.

# Generating Standard Reports

Standard reports are already designed for you, offering numerous choices regarding the information you can include. You do little more than click a few buttons to generate them. Essentially, you select a report category, choose a specific report, and print it. If the plain-vanilla version of a report isn't quite right, you can modify standard reports in a variety of ways.

TIP

You can also print any view in Project: Simply press Ctrl+P, and then click Print. The entire project is printed in whichever view is onscreen at the time. Or you can choose File ⇨ Print. In the Print pane in the Backstage view you can choose to print only certain pages of the project or only a specific date range from the timescale. Any filter or grouping that you've applied is shown in the printed document.

## What's available on the Report tab

On the Report tab, Project has four standard report categories: Dashboards, Resources, Costs, and In Progress. Each category contains several predesigned reports (as you can see on the Report tab of the Ribbon, shown in Figure 18-1), for a total 16 standard reports. There is also a set of reports for Task Boards. I cover Task Boards in Chapters 19 and 20, so in this chapter, I focus on the 16 standard reports.

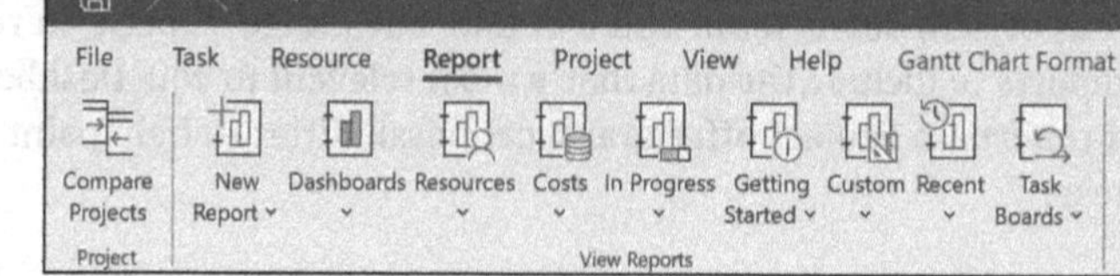


**FIGURE 18-1:** The Report tab.

© John Wiley & Sons, Inc.

Standard reports vary in content, format (for example, a table, chart, or a comparison report), and sometimes page orientation (landscape or portrait). You can edit a report to change its name, the period it covers, the table of information it's based on, and the filters applied to it. You can also sort information as you generate the report and add formatting, such as themes, images, and shapes.

## Dashboard reports

Dashboard reports are a cool way to provide an overview. Project has five standard dashboard reports: Burndown, Cost Overview, Project Overview, Upcoming Tasks, and Work Overview.

To see a dashboard report, simply click the Report tab, click the Dashboards down arrow, and select the report you want. Figure 18-2 shows an example of a Project Overview dashboard report.

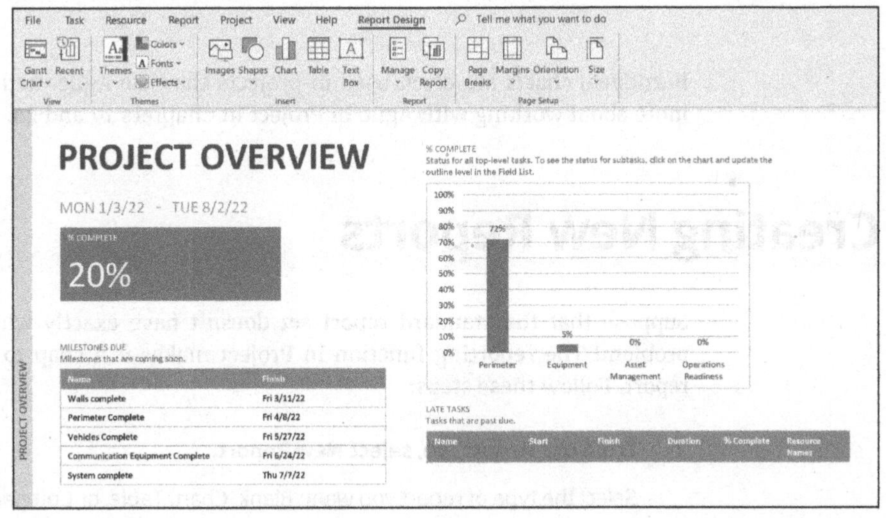


**FIGURE 18-2:** Project Overview report.



A dashboard report you may not be familiar with is a Burndown chart. You use it to review the amount of work remaining compared to the baseline work remaining and the number of tasks remaining compared to the baseline remaining tasks. On the Burndown chart, you can see the rate of work compared with the planned rate of work to determine whether you're accomplishing work at the same rate as planned. If the number of tasks remaining is higher than the baseline rate, you're in jeopardy of delivering late. Figure 18-3 shows a Burndown chart.

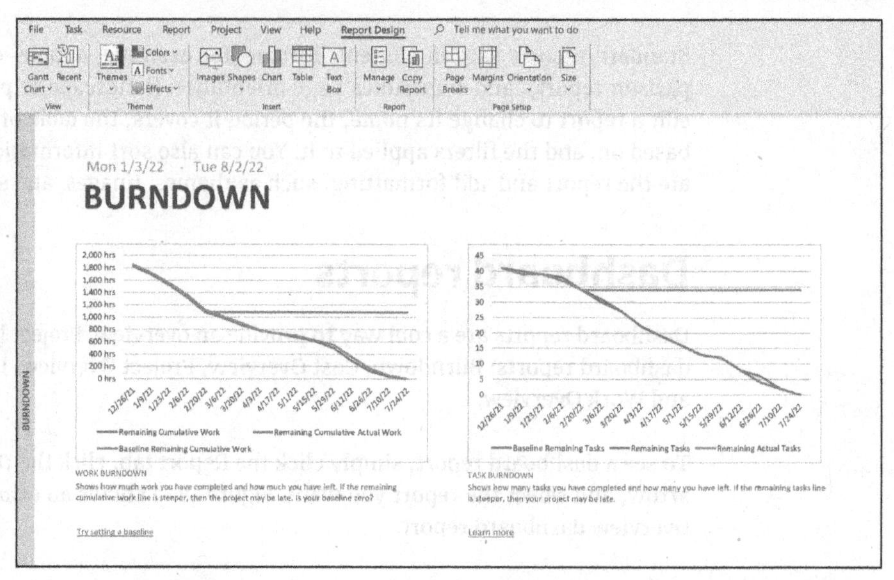


**FIGURE 18-3:** Burndown report.

*© John Wiley & Sons, Inc.*

Burndown charts are often used in projects that use Agile methodologies. I talk more about working with Agile in Project in Chapters 19 and 20.

# Creating New Reports

Suppose that the standard report set doesn't have exactly what you want. No problem! The reporting function in Project makes it a snap to create a custom report. Follow these steps:

1. **From the Report tab, select New Report.**

   Select the type of report you want: Blank, Chart, Table, or Comparison.

2. **When the Report Name dialog box opens, enter a name for the report and click OK.**

   If you choose a Chart, Table, or Comparison report, a split window opens, as shown in Figure 18-4; the left side holds the report, and the right side holds a Field List.

3. **In the Select Category drop-down list, choose among Time, ID, Name, Resource Names, and Unique ID.**

4. **From the Select Fields expandable list, select the check box next to the fields you want from the various categories (cost, duration, number, or work).**

5. **If you want to apply a filter, such as incomplete tasks or active tasks, select one from the Filter drop-down list.**

6. **If you want to group information, select the appropriate grouping from the Group By drop-down list.**

7. **Select at which level you want to see information reported using the Outline Level drop-down list — for example, Level 1 (the entire project), Level 2, or Level 3.**

8. **Choose how to sort the data.**

   This is the order in which you want to present the data based on the fields you selected in Step 4.

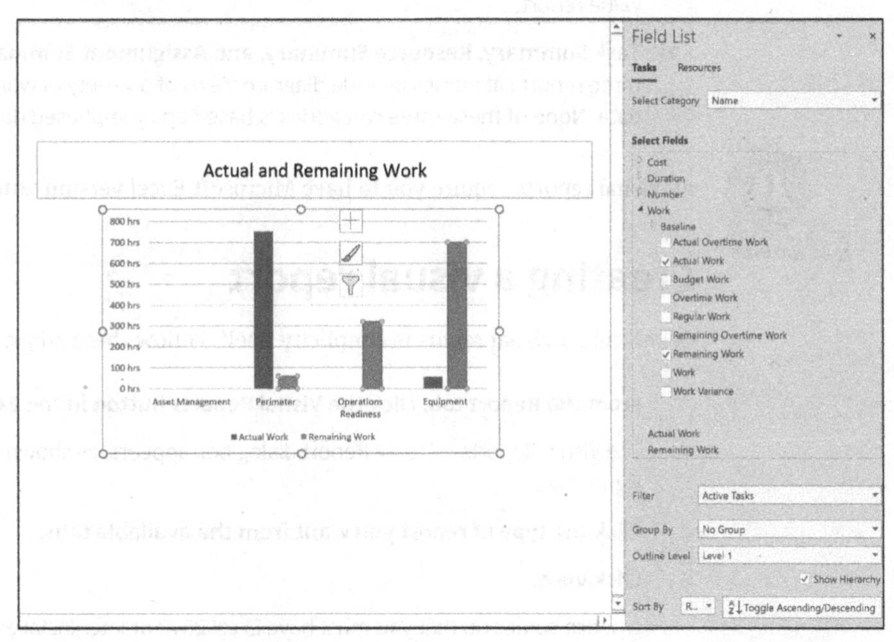


**FIGURE 18-4:** Building a custom report.



After you create a new report, you can access the report in the Custom category in the Reports dialog box or by choosing Report⇨View Reports⇨Custom and then clicking the report's name.

## Gaining a new perspective on data with visual reports

Project offers six categories of visual reports and some custom reports that you can build yourself. Some visual reports are based on *time-phased data* (data distributed over time, such as allocations of resource time or costs), and some aren't. Report categories are described in this list:

- **Task Usage:** Based on time-phased data for tasks, this report category lets you peek at cash flow over time using Excel.
- **Resource Usage:** Based on time-phased resource data, this type of report includes resource costs, resource availability, and a resource work summary.
- **Assignment Usage:** Also based on time-phased data, this category provides information in areas such as baseline versus actual costs, baseline versus actual work, budget cost, and budget work reports. It also provides an earned value report.
- **Task Summary, Resource Summary, and Assignment Summary:** These three report categories provide diagram views of a variety of work and cost data. None of these three categories is based on time-phased data.

WARNING

All visual reports require you to have Microsoft Excel version 2010 or later.

## Creating a visual report

Generating a visual report is simplicity itself. Follow these steps:

1. **From the Report tab, click the Visual Reports button in the Export group.**

   The Visual Reports – Create Report dialog box appears, as shown in Figure 18-5.

2. **Click the type of report you want from the available tabs.**
3. **Click View.**

   A notice comes up that you must have Excel 2010 or later installed on your computer. Just click OK to eliminate the box.

TIP

To customize a visual report, you need to know about pivot tables. Using pivot tables, you can view data from a variety of perspectives beyond the standard Project report capabilities. Pivot tables offer perspectives that are especially useful for data analysis. Because a discussion of pivot tables in those products is beyond the scope of this book, I heartily recommend *Excel For Dummies* by Greg Harvey (Wiley) or *Excel Data Analysis For Dummies* by Paul McFedries (Wiley).

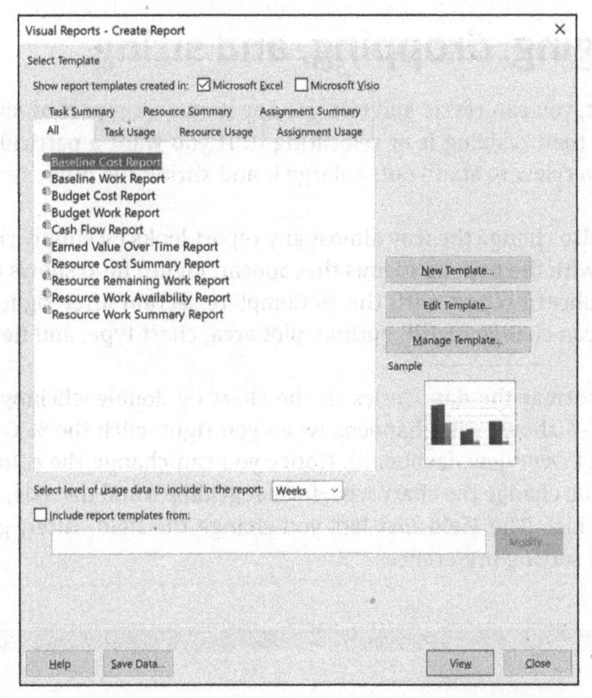


**FIGURE 18-5:** The Visual Reports dialog box.

*© John Wiley & Sons, Inc.*

TIP

You can modify the Visual Reports templates or create your own templates by using the Edit Template and New Template buttons in the Visual Reports dialog box. When you edit a template, you add fields to it or remove fields from it. Create a new template by selecting the available and custom fields you want to include.

# Fine-Tuning a Report

After you have lots of data available to you in Project reports, it would be a shame not to customize the presentation to maximize its impact. Project helps you create reports that not only show off the information, but also make the information look good!

As soon as you start creating a report, Project displays the Report Design contextual tab. From this tab, you can work with themes; insert images, shapes, charts, tables, and text boxes; and work with Page Setup in the print function. Before getting into all the wonderful ways you can make reports attractive, I cover some of the easy ways you can manipulate the information in the reports.

## Dragging, dropping, and sizing

In Project, you can resize and relocate any item on a report by simply clicking the item and then resizing it or relocating it. If you want a particular section of the project overview to stand out, enlarge it and shrink the other sections.

You can also change the way almost any report looks by simply right-clicking and working with the pop-up menus that appear. Figure 18-6 shows the Project Overview dashboard report with the % Complete section after right-clicking. Notice that you can change its fill, outline, plot area, chart type, and fields.

You can format the data series in the chart by double-clicking a chart section. Figure 18-6 shows what happens when you right-click the % Complete chart on the Project Overview dashboard. Notice you can change the colors of the fill and outline. You change the chart type, format gridlines and the axis, and show or hide the Field List. The Field List lets you change the field, filter, grouping, outline level, and sorting preferences.

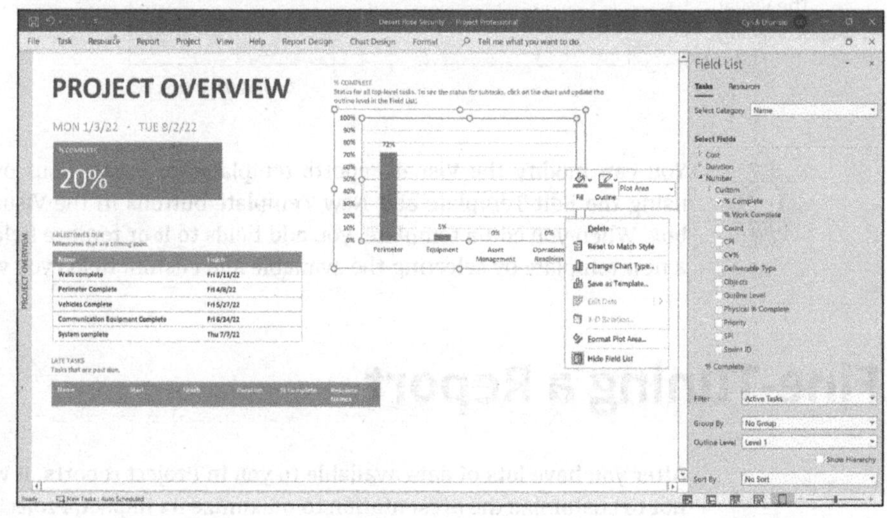


**FIGURE 18-6:** Changing the report layout and format.

© John Wiley & Sons, Inc.

Along the right side of the chart are different floating buttons: one for chart elements, one for style and color options, and a filter.

- The Chart Elements button lets you select elements such as Axis Titles, Data Labels, Gridlines, Legend, Trendline, and other options.
- The Format button provides lots of options for the Chart style and the Chart Color pallet.

## Looking good!

If you have no creative flair, don't worry — Project does. You can work with the themes that Project provides to present all the reports with flair. Choose Report Design⇨Themes⇨Themes to display the Themes gallery. You see 30 different themes that you can apply to reports. Themes include color, type, font, and effect. Play around with a few to find one you like.

If none of the themes works for you, look around on the Colors drop-down list, Fonts drop-down list, and Effects drop-down list until you create *just* the look you're hoping for.

To customize the reports even more, you can add images such as a team logo, shapes, and text boxes. Figure 18-7 shows a portion of a report with a text box and shapes added.

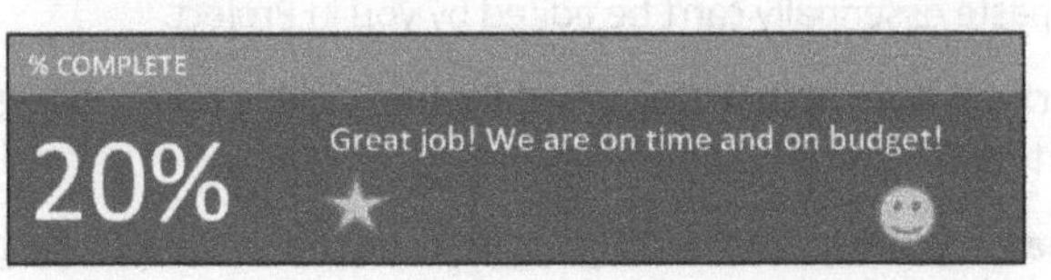


**FIGURE 18-7:** Customizing with shapes and text boxes.

© John Wiley & Sons, Inc.

REMEMBER

Experimenting with customization options can be fun, but as with any other business document, the goal in layout and formatting is presenting information clearly.

Keep these points in mind when you're formatting Project text:

- **Font:** Choose simple sans serif fonts, such as Calibri or Arial. If you're publishing the project on the web, consider using Verdana, which is a font created for readability online.
- **Color:** You have to consider factors such as whether the printout will be in color or black and white, whether the use of too many colors will become confusing for the reader, and whether certain colors (such as yellow) will be difficult to read.
- **Font size:** Use a font size that's readable, without making it so big that the taskbar labels become too crowded.
- **Effects:** Avoid text effects that can make certain text difficult to read. Use these effects only to call attention to a few elements of the project.

## Spiffing Things Up

**These days, image is everything. You and the project may be judged to a certain extent by how professional the printed information looks. Even if the project is over budget and four months behind, making reports or other printouts look good can make delivering bad news easier to do. You can make the project presentation look good by adding graphics.**

**Wouldn't the company logo look spiffy in the header of the report? Or a picture of the new product box in Gantt Chart view of the New Product Launch project?**

**Graphics can add visual information or make the plan just plain look nicer. You can insert graphics in the project file by using one of these three methods:**

- **Cut and paste a graphic from another file.** A graphical image that you cut and paste essentially can't be edited by you in Project.
- **Insert a link to an existing graphics file.** Linking takes up less room than inserting the image.
- **Embed a graphic.** Embedding lets you edit the content of the graphic in Project, using the tools of an image program such as Paint.

**You can't add graphics willy-nilly, however. You can add graphics in only a few places: the chart pane of any Gantt Chart view; a task note; a resource note; or a header, footer, or legend that's used in reports or printouts of views.**

**For example, you may put pictures of resources in the Resource Note field so that you can remember who's who, or you may include a photo of the corporate headquarters in the header of the report.**

**When you paste or embed graphics, the Project file's size can swell like a sponge in a pail of water. If you're thinking of using a lot of graphics, don't let them detract from the main information in printouts. Or try linking to them instead of inserting them into the file.**

**If you want to insert an existing graphics file into a Notes box, follow these steps:**

1. **Double-click a task in the Gantt Chart view. Then select the Notes tab in the Task Information or Resource Information dialog box.**
2. **Click the Insert Object button.**

   The Insert Object dialog box, shown in Figure 18-8, appears. The arrow points to the Insert Object button.

3. **Choose the Create from File radio button.**
4. **Click the Browse button to locate the file.**
5. **To insert to the file, click the Insert button and then click the OK button.**

   You may see a warning before you insert an image. If you want to continue, click Yes.
6. **To link to the file, select the Link check box next to the Browse button.**

   If you don't select this check box, the object is embedded in the file.
7. **To insert the object as an icon, select the Display as Icon check box.**

   TIP

   When you display the object as an icon, someone viewing the project on a computer can click the icon to view the picture.
8. **Click the OK button.**

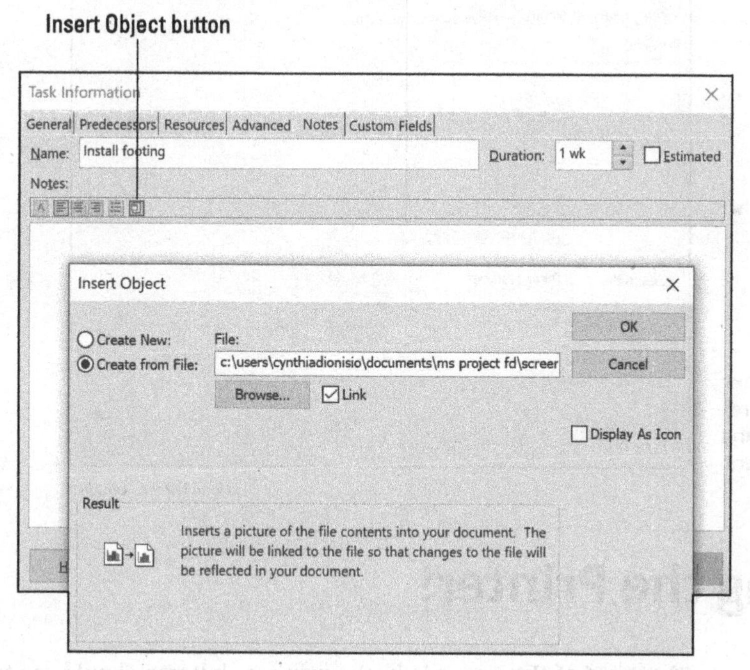


**FIGURE 18-8:** Inserting graphics.

© John Wiley & Sons, Inc.

If you want to insert an existing graphics file in a header or footer, follow these steps:

1. **Choose File ⇨ Print.**
2. **Click the Page Setup link below the Settings list.**

3. **Go to the Header or Footer tab in the Page Setup dialog box.**
4. **Click the Insert Picture button, as shown in Figure 18-9.**

   The Insert Picture dialog box appears.
5. **Navigate to the folder that holds the file to insert.**
6. **Click to select the file.**
7. **Click the Insert button.**

   The image appears in the header or footer.
8. **Click OK.**

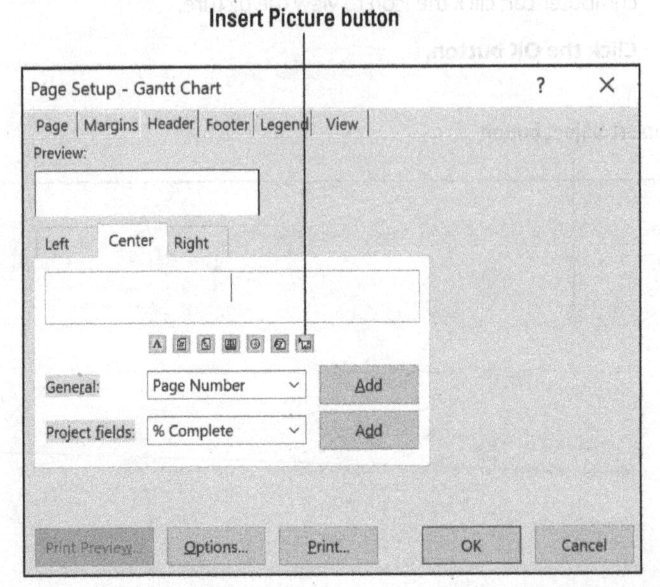


**FIGURE 18-9:** Inserting pictures in headers and footers.

© John Wiley & Sons, Inc.

# Calling the Printer!

The proof of the report is in the printing, but you should see to several adjustments before you click the button to print. In Project, it's not only margins or page orientation that you need to set (although you have to set them, too); you can also add useful information to headers and footers and set legends to help readers understand the many bars, diamonds, and other graphical elements that many Project views and reports display.

## Working with Page Setup

The Page Setup dialog box can be used to control printouts of any displayed Project view. You can open this dialog box by choosing File⇨Print⇨Page Setup. The Page Setup dialog box, as shown in Figure 18-10, contains six tabs.

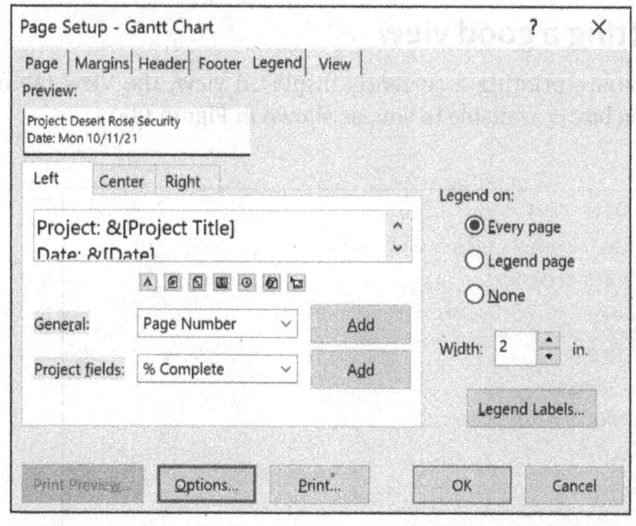


**FIGURE 18-10:** Page setup.

© John Wiley & Sons, Inc.

The information in the Page, Margins, Header, and Footer tabs is standard Microsoft content. I won't bore you by going over it. On the other hand, information in the Legend and View tabs is more specific to Project.

### Working with a legend

A *legend* acts as a guide to the meanings of various chart elements. It is similar to the Header and Footer tabs, but the legend is generated automatically. You can see the Legend tab in Figure 18-10. Other differences from the Header and Footer tabs include:

- You can print the legend on every page or on a separate legend page, or you can decide not to print a legend.
- You can establish the width of the text area of the legend (the area where you can insert elements such as the page number or the date).
- You can edit the text font, style, size, and color of the legend by clicking the Legend Labels button.

Figure 18-11 shows the legend for the Desert Rose project.

Project: Desert Rose Security
Date: Mon 10/11/21

Task
Split
Milestone
Summary
Project Summary
Inactive Task
Inactive Milestone
Inactive Summary
Manual Task
Duration-only
Manual Summary Rollup
Manual Summary
Start-only
Finish-only
External Tasks
External Milestone
Deadline
Critical
Critical Split
Progress
Manual Progress

**FIGURE 18-11:** Legend for the Desert Rose project.

© *John Wiley & Sons, Inc.*

## Getting a good view

If you are printing a currently displayed view, the View tab of the Page Setup dialog box is available to you, as shown in Figure 18-12.

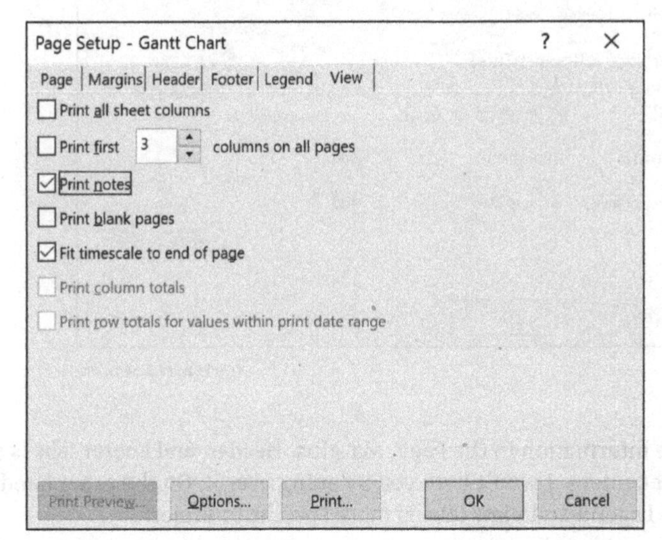


**FIGURE 18-12:** The View tab.

© *John Wiley & Sons, Inc.*

You can make these settings on the View tab:

- **Print All Sheet Columns:** Prints every sheet column in the view, regardless of whether it's visible onscreen. With this check box unchecked, only the columns that show in the view are printed.
- **Print First # Columns on All Pages:** Lets you specify a specific number of columns to print, such as the information, WBS, and Task columns.
- **Print Notes:** Prints every task, resource, and assignment note. These items are printed on a separate Notes page.
- **Print Blank Pages:** Lets you print blank pages. For example, select this check box to print a page that represents a time in the project when no tasks are occurring. If you want a smaller number of pages in the printout, ignore this setting.

- **Fit Timescale to End of Page:** This nifty little setting scales the timescale to fit more of the project on the page.
- **Print Column Totals:** Adds a column that contains column totals. It pertains to printouts of Usage views. If you aren't printing a Usage view, this option is not available.
- **Print Row Totals for Values within Print Date Range:** Adds a column that contains row totals. It pertains to printouts of Usage views. If you aren't printing a Usage view, this option is not available.

## Getting a preview

Although software print previews aren't quite as exciting as movie previews, they help you make everything look right before you print the project. When you choose a report or choose File⇨Print to print a view, the preview automatically appears in the Backstage. Figure 18-13 has a box around the buttons in the lower-right corner of the preview that enable you to modify the preview so that you can work with the settings more effectively.

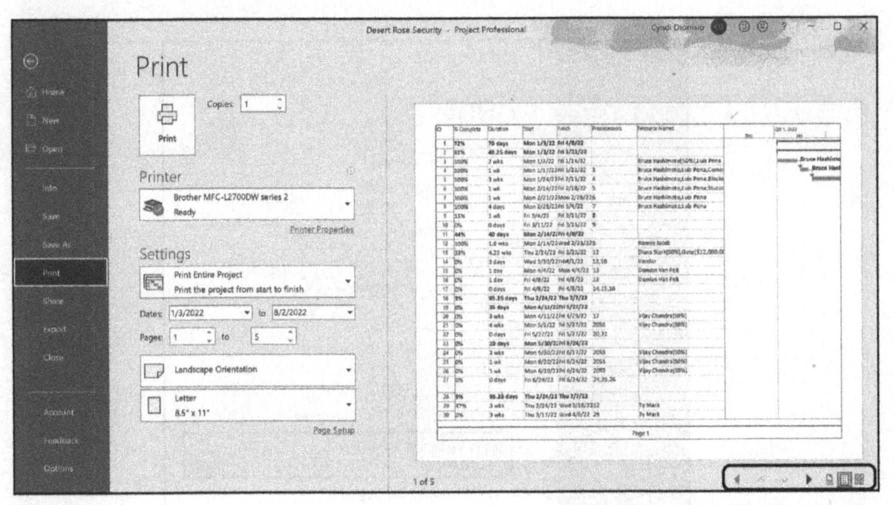


FIGURE 18-13: Backstage Print view.

© John Wiley & Sons, Inc.

You can use the buttons in the Backstage preview to:

- Move around the pages of the report by using the first four arrow buttons.
- View more detail by clicking the Actual Size button and then using the vertical and horizontal scroll bars to zero in on the section you want to see.

» Focus on a single page or all pages in the report by using the One Page and Multiple Pages buttons.

TIP

Always preview the document before printing it to review the number of pages being printed. One click or one instance of zooming in or out can easily double — or cut in half — the number of pages being sent to the printer. You can easily waste paper if you're not careful.

## Finalizing your print options

Before you send your masterpiece to the printer, pay attention to the information on the left side of the Backstage Print view. Here you set the number of copies you want to print and select the printer you will use. This is also where you indicate if you want to see the entire project, specific dates, or specific pages. In addition, you can specify the orientation and paper size.

If you click the arrow in the first drop-down list under the Settings heading, you will see the printing options shown in Figure 18-14, including the option to print notes, all sheet columns, or the left column of pages only.

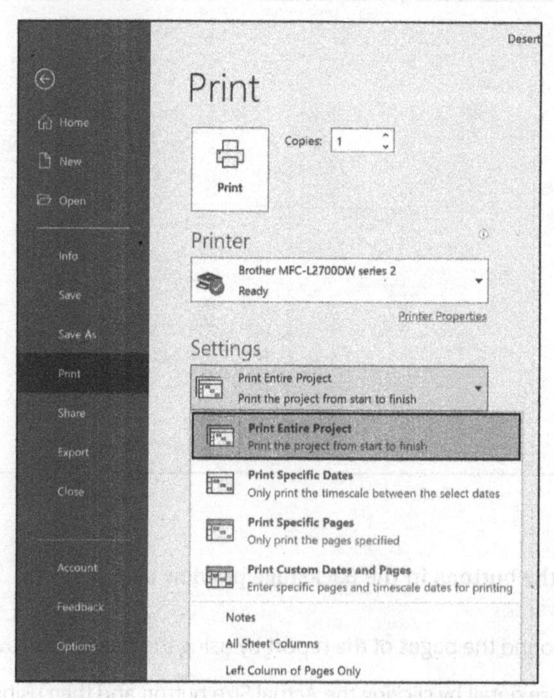


**FIGURE 18-14:** Print settings.

© John Wiley & Sons, Inc.

After you choose all the settings and you're ready to print, simply click the big Print button in the upper-left corner.

# Working on the Timeline

The Timeline gives sort of a 10,000-foot overview of the project schedule. It shows a graphical timeline spanning from the project's start date to its end date, and it highlights the period shown in the main view, such as the timeframe shown on the Gantt chart. You can toggle the Timeline on and off as needed. To do so, click the View tab on the Ribbon and check or uncheck the Timeline check box in the Split View group.

The Timeline even has its own Ribbon tab, the Timeline Format contextual tab, as shown in Figure 18-15.

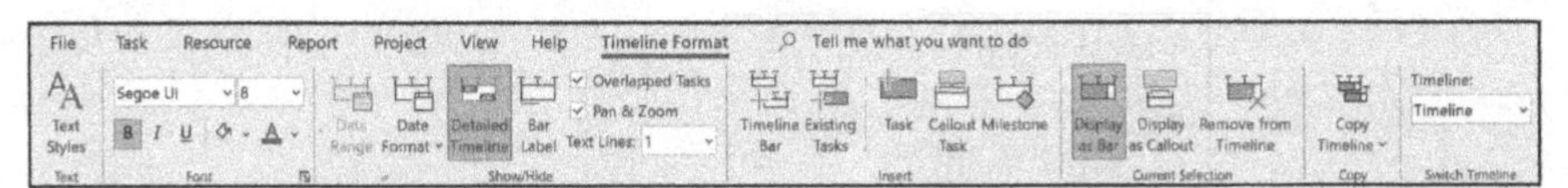


**FIGURE 18-15:** The Timeline Format contextual tab.

© John Wiley & Sons, Inc.

From the Timeline Format contextual tab, you can format text and dates, format the display of tasks on the timeline, and copy the Timeline in various presentation formats.

## Adding tasks to the Timeline

You can add summary tasks, detailed tasks, and milestones to the Timeline. For example, the Timeline shown in Figure 18-16 shows the summary tasks and milestones along the bottom.

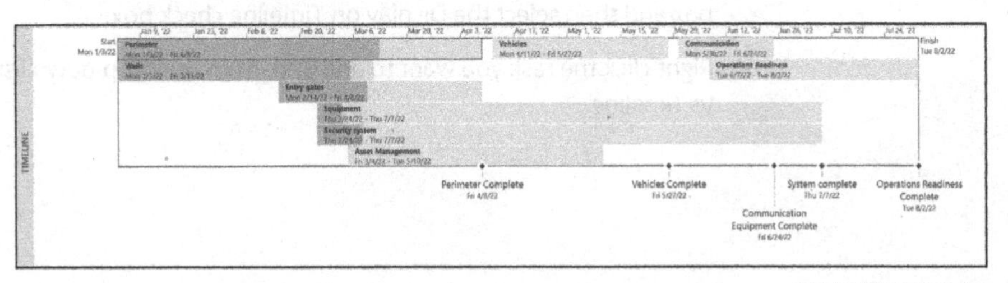


**FIGURE 18-16:** The Timeline with summary tasks and milestones.

© John Wiley & Sons, Inc.

You can add information to the Timeline in a number of ways. To add several tasks at a time, follow these simple steps:

1. **Show the Timeline by checking the Timeline box in the Split View group on the View tab.**

   The timeline will be visible in the chart area, and the Timeline Format contextual tab will appear.

2. **Click the Timeline Format tab.**
3. **Click Existing Tasks in the Insert group.**

   The Add Tasks to Timeline dialog box opens, as shown in Figure 18-17, showing an outline of all tasks in the project. You can select as many as you like. They appear on the Timeline as individual tasks, summary tasks, or milestones, depending on how they're entered in Project.

4. **Click OK.**

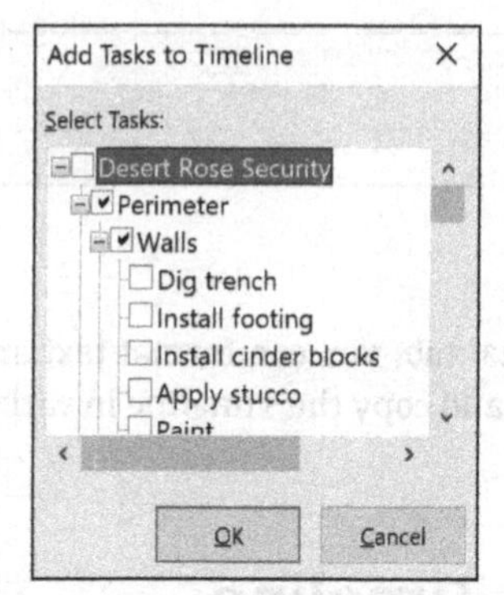


**FIGURE 18-17:** Adding tasks to the Timeline.

*© John Wiley & Sons, Inc.*

There are two additional ways to add tasks to the Timeline:

- Click the Task button in the Insert group to open the Task Information dialog box and then select the Display on Timeline check box.
- Right-click the task you want to add and, from the drop-down list, choose Add to Timeline.

## Customizing the Timeline

Project lets you customize the Timeline to make it truly fit your needs. If you want to change the way a task appears on the Timeline, for example, follow these steps:

1. **Click to select the task.**
2. **In the Current Selection group, select either Display as Bar to show a bar that indicates the duration or Display as Callout to create a callout box that holds the task information. If you want to remove a task from the Timeline, select Remove from Timeline.**
3. **Reposition the callout box to improve readability, if you have several tasks close together.**

You can also change the date format from the Show/Hide group on the Format tab in the Timeline Format contextual tab group. Simply click the down arrow and select the way you want to see the date displayed on the Timeline.

The Timeline can be displayed with details about tasks, such as the task name and dates, or without the detail and just milestones and bars that represent tasks, but no names associated with the tasks. If you want a detailed view, click the Detailed Timeline button in the Show/Hide group.

## Copying the Timeline

Because that vendor you're trying to coordinate with doesn't use Project, you need to share the graphical Timeline information via another file format. Project builds in the special capability to take a snapshot of the Timeline so that you can paste it into an email message; a presentation; or another program, such as PowerPoint.

Follow these steps to copy and use the Timeline:

1. **Create the email message or document where you want to use the Timeline image.**
2. **Choose Timeline Format ⇨ Copy ⇨ Copy Timeline, and then select a format.**

   The available format choices are For E-Mail, For Presentation, and Full Size.
3. **In the program where you want to insert the Timeline, position the cursor where you want the Timeline image to appear.**
4. **Press Ctrl+V.**

   The image appears in the message or document.

# 5 Working with Sprints Projects

IN THIS PART . . .

Set up a Sprints Project.

Work with filters, tables, groups, and sorting to manage Sprints Projects.

Learn from your projects.

IN THIS CHAPTER

» Creating a Sprints Project

» Organizing sprints

» Agile views

» Sprint planning

» Customizing Sprints Projects

# Chapter 19
# Setting Up a Sprints Project

The way we manage projects has progressed to include approaches for both plan-driven projects with defined scope and adaptive projects where the scope can evolve along the way. Project can support both approaches. In fact, it is so versatile that it can support hybrid projects that use plan-driven and adaptive approaches on the same project.

In this chapter you learn how to set up an adaptive project using sprints. You learn about several different views that Project has to help you plan Sprints Projects. It also covers customizing the content you see on the Task Board. If you are new to Sprints Projects, some of the terminology and the methods used to plan and track projects will be unfamiliar, but the chapter takes it one step at a time and eases you into this whole new way to work on projects.

## Creating a Sprints Project

If you are new to adaptive or Agile projects, you may be asking yourself, "What is a sprint?" and "What is a Sprints Project?" A *sprint* is a timebox or time interval to accomplish work. Sprints usually last one week, two weeks, or four weeks. Sprints Projects keep the sprint durations the same throughout the project.

A fundamental difference between a Sprints Project and a plan-driven or Waterfall Project is that, with a Waterfall Project, the scope is fixed and you estimate how long it will take to accomplish that scope. In a Sprints Project, there are fixed timeboxes and you estimate how much work you can accomplish in a timebox.

You create a Sprints Project the same way you do any other project. As shown in Figure 19-1, from the New pane in Backstage view, simply choose Sprints Project.

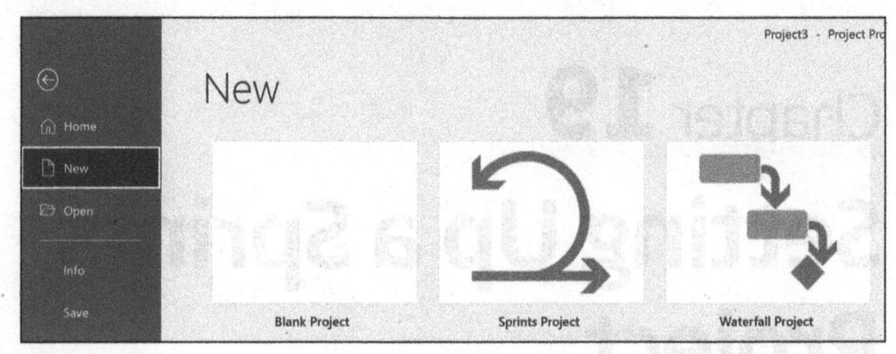


**FIGURE 19-1:** Creating a new Sprints Project.

© John Wiley & Sons, Inc.

When you create a Sprints Project, you will see a Sprints tab on the Ribbon. Figure 19-2 shows the Sprints tab.

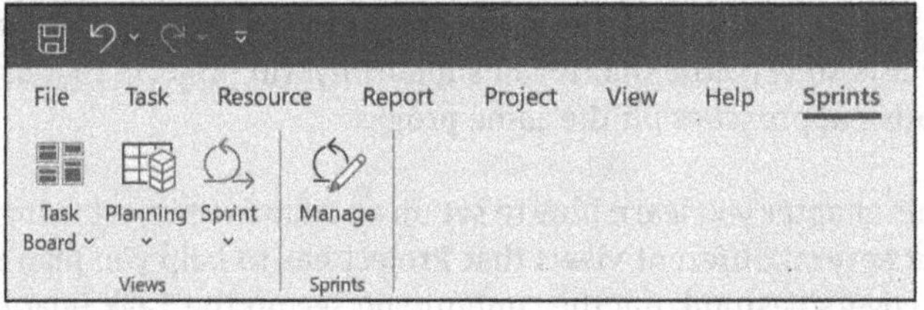


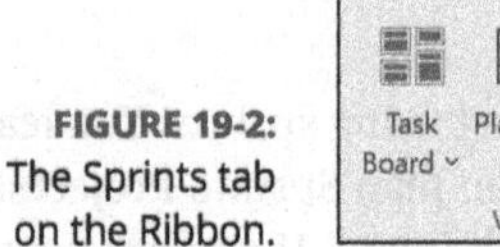

**FIGURE 19-2:** The Sprints tab on the Ribbon.

© John Wiley & Sons, Inc.

To set up your project, you start by setting up your sprints.

1. **On the Sprints tab, click Manage in the Sprints group.**

   The Manage Sprints dialog box appears, as shown in Figure 19-3. By default, it shows No Sprint, Sprint 1, Sprint 2, and Sprint 3.

2. **Name your sprints.**

   If you want to have more interesting names than Sprint 1, Sprint 2, and Sprint 3, click in the Name column and type a brief name that describes the work that will be done in that sprint, such as Design, Build Database, or Develop Reports.

3. **Set your sprint length.**

   You set the length (duration) by typing it in the Length column or using the spinner arrows to set the length. The default duration is 2 weeks; however, as you can see in Figure 19-3, the sprints were changed to 1 week.

4. **Set the start date.**

   Once you set the start date for the first sprint, Project automatically fills in the finish date and the start dates for the following sprints based on the sprint duration. Note that in a Sprints Project, all sprints are back-to-back. In other words, one sprint has to finish before the next one begins.

5. **Add more sprints.**

   If you want to add more sprints, go to the Add Sprint section at the bottom of the Manage Sprints dialog box. You will see that the start date is populated to start after the previous sprint is finished. Just set the duration by typing it in or using the up or down arrows. For Sprints Projects, the durations show up in elapsed weeks. An elapsed (or calendar) week is shown as ew (such as 1 ew, 2 ew, and so on). *Ew* stands for elapsed calendar weeks, which include weekends in the duration.

   If you don't know how many sprints you need, you can always come back and add more later. Once you enter information about a new sprint, click the Add Sprint button. The new sprint will be added to the list of sprints.

6. **When you are done adding sprints, click OK.**

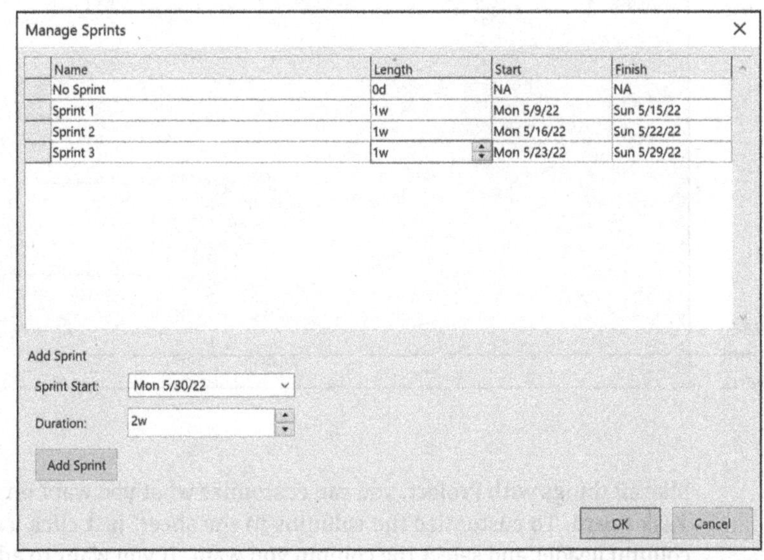


**FIGURE 19-3:** The Manage Sprints dialog box.

TIP

**The No Sprint field in the top row of the Manage Sprint box is used to hold tasks that haven't been assigned to a sprint yet.**

# Enjoying a Whole New View

**Sprints Projects use different views than plan-driven projects. Rather than Gantt charts, network diagrams, and calendars, Sprints Projects use views for backlogs and sprints. There are two ways to work with Sprints Projects — using sheets or using boards. A sheet is just like the sheet in a plan-driven project, except it has different columns. Figure 19-4 shows a Sprint Planning sheet and Figure 19-5 shows a Sprint Planning Board.**

| | ⓘ | Sprint | Name | Work | Board Status | Resource Names | Task Summary Name | Deadline | Show on Board |
|---|---|---|---|---|---|---|---|---|---|
| 2 | | Sprint 1 | Setup | 0 hrs | Prior Sprints | | | NA | Yes |
| 3 | | Sprint 2 | Architecture | 0 hrs | Prior Sprints | | | NA | Yes |
| 4 | | Sprint 2 | Requirements | 0 hrs | Prior Sprints | | | NA | Yes |
| 1 | | Sprint 3 | Build database | 0 hrs | In progress | | | NA | Yes |
| 5 | | Sprint 3 | Setup users | 0 hrs | Next up | | | NA | Yes |
| 6 | | Sprint 3 | Setup permissions | 0 hrs | Next up | | | NA | Yes |
| 7 | | Sprint 4 | Load vehicle assets | 0 hrs | Not Started | | | NA | Yes |
| 8 | | Sprint 4 | Load vehicle vendors | 0 hrs | Not Started | | | NA | Yes |

**FIGURE 19-4:** The Sprint Planning sheet.

*© John Wiley & Sons, Inc.*

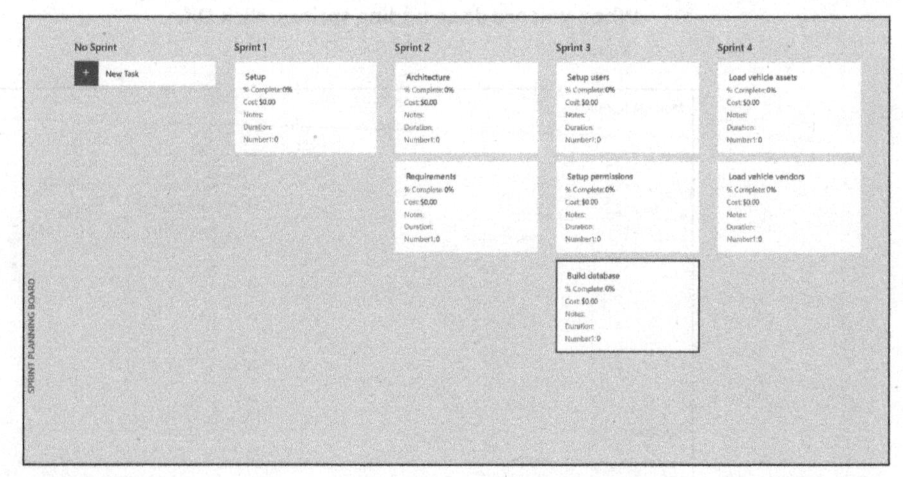


**FIGURE 19-5:** The Sprint Planning Board.

*© John Wiley & Sons, Inc.*

**Like all things with Project, you can customize what you want on the sheet and the Task Board. To customize the columns in the sheet, just click a down arrow on a column header and select the column you want. If you want to add columns to the existing set, go to the Format tab and choose the Columns group⇨Insert Column.**

Customizing the information on the sprints board and the cards on the board is covered later in this chapter.

TIP

A card is a like an electronic index card that holds information about a task.

There are two ways to see the different views for a Sprints Project. The first way is from the Sprints tab, in the Views group. Click the down arrow by the Task Board. You will see the Task Board and Task Board sheet.

For a wider range of views, choose the View tab➪Task Views➪Task Board. Click the down arrow on the Task Board button to see views for the Backlog Board, Current Sprint Board, Sprint Board, Sprint Planning Board, and Task Board. If you want even more options, choose the View tab➪Task Views➪More Views. You can scroll through the different views and select the one you want. Figure 19-6 shows a few views from the More Views option.

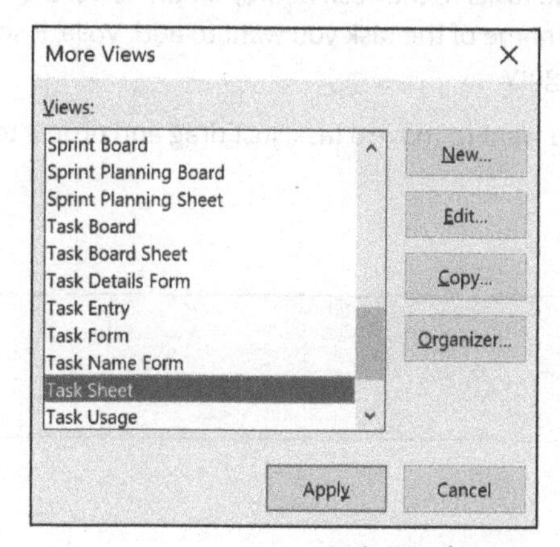


**FIGURE 19-6:** Sprint and Task views.

© John Wiley & Sons, Inc.

# The Task Board and Task Board sheet

The Task Board is where you set up the way you want to track your work. It is a virtual Kanban board. At the start of the project it will look like Figure 19-7. But with a few tweaks you can customize it to meet the needs of your project. Consider the following options:

» **Customizing columns:** By default the columns are set up as Not Started, Next up, In Progress, and Done. If these categories work for you, you are set.

- **Adding a column:** Let's say you want to add a column for Testing and another for Previously Done so you can see all the work that has been completed. All you have to do is click Add New Column on the far right and type the name of the column you want to add.
- **Moving a column:** To move a column, right-click the column and choose Move Right or Move Left.
- **Renaming a column:** To rename a column, right-click and choose Rename. Then type the name you want.
- **Deleting a column:** To delete a column, right-click and choose Delete.
- **Setting % Complete:** This button lets you indicate the percentage complete of the work in a column that has work in progress. You can show or hide the Set % Complete information by going to the Task Board Format contextual tab and selecting or deselecting the Show % Complete Mapping check box.
- **Adding tasks:** To add tasks to the Task Board, simply select the + New Task button and type the name of the task you want to add. Voila, it shows up in the Not Started category.
- **Moving tasks:** If you want to move a task, just drag and drop it to the appropriate sprint.

Not Started
SET % COMPLETE
+ New Task
Next up
SET % COMPLETE
In progress
SET % COMPLETE
Done
% COMPLETE: 100
Add New Column

**FIGURE 19-7:** Task Board view.

© John Wiley & Sons, Inc.

The Task Board sheet has information from the Task Board, plus the Work, Resources, Task Summary, and Deadline columns.

## The Sprint Planning Board and Sprint Planning sheet

The Sprint Planning Board shows which work is planned for which sprint. This allows for high-level planning and ensuring that you have the right resources available when you need them. The Sprint Planning Board is shown in Figure 19-5.

The Sprint Planning sheet reflects information you entered elsewhere. For example, in the Sprint column there is a drop-down menu that has the name of the sprints you defined in the Manage Sprints dialog box. The Board Status column has a drop-down menu with categories of work you identified in the Task Board view.

You can add or hide columns in this view just like any other view. For example, you may not want to track the Task Summary name, but you may want to show the cost associated with a task. One handy feature in this sheet is the Show on Board column. If you want the task to show up on a Task Board, choose Yes. If not, choose No. The Sprint Planning sheet is shown in Figure 19-4.

## The Current Sprint Board and Current Sprint sheet

These views work with the current sprint only. They reflect the information you entered in the Manage Sprints dialog box (the name and dates of the sprints) and in the Task Board (categories of work).

## The Backlog Board and the Backlog sheet

A *backlog* is a list of work to be done. In Sprints Projects, the product owner prioritizes the work on the backlog so the team works on the most important items first. Therefore, the order of the work in the Not Started column can be changed to reflect the most important work. This also allows you to add new features, functions, and tasks.

The backlog views show all the work in the project and the current status of each task. These are good views for getting an overview of the entire project, how much work has been done, and how much is left to do.

REMEMBER

If you edit information in one view, Project updates the information in all the other views.

Table 19-1 provides a brief summary of views for Sprints Projects.

**TABLE 19-1 Sprints Project Views**

| Variable | Description |
|---|---|
| Task Board | Tracks the status of work. It has categories of work in progress such as Not Started, Next up, In Progress, and Done. These categories can be customized to meet the needs of your project. |
| Task Board Sheet | Contains the same information as the Sprint Planning sheet, except for the Show on Board column. |
| Sprint Planning Board | Used to identify and name sprints, then assign work to specific sprints. |
| Sprint Planning Sheet | Contains information on tasks such as the Sprint, Task Name, Work (the amount of work the task is estimated to take), Board Status (Not Started, Next Up, and so on), Resource Names, Task Summary Name, Deadline, and Show on Board. You can add or delete columns to track the information you want. |
| Current Sprint Board | Shows the task cards that are active for the current sprint and which category they are in (Not Started, Next Up, and so on). |
| Current Sprint Sheet | Shows task information for the current sprint in a sheet view. |
| Backlog Board | Shows the status of all project work. Work that hasn't started is in the backlog. A backlog is a list of work to be done. The work in the upcoming sprint is in the Next Up column. Work in the current sprint is in the In Progress column. Work that is done is in the Done column. |
| Backlog Sheet | Shows the same information as the Backlog Board, but includes columns for Work (the estimated time it will take to accomplish a task), Task Summary, and Deadline. |

# Adding Information to Tasks

To get a more complete view of task information, you can add columns in the sheet views. To see that information on Board views, you can customize the cards you see in those views. To customize the cards, you have to be in a Board view, such as the Task Board. Once you are in a Board view, you see the Task Board Format contextual tab next to the Sprints tab, as shown in Figure 19-8. This tab allows you to show or hide the % Complete Mapping. More importantly, it allows you to customize the information you see on the task cards.

Figure 19-9 shows the Customize Task Board Cards dialog box. You can choose to show the task ID, resources, and a checkmark for work that is complete. You can also add up to five fields of information on your cards. Just click the arrow on the Select Field box to see all the different options. Figure 19-10 shows cards where the Show Resources check box and the Show 100% Complete check box are checked. The cards also have fields for Priority, Cost, % Complete, and Notes.

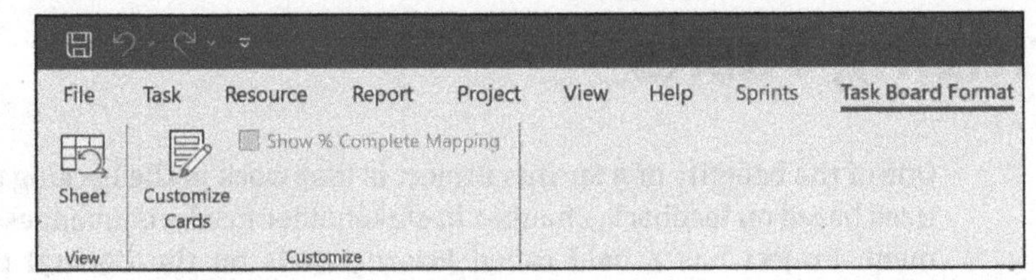


**FIGURE 19-8:** The Task Board Format contextual tab.

*© John Wiley & Sons, Inc.*

Customize Task Board Cards

Base Fields

Show Task ID

Show Resources

Show checkmark when 100% complete

Additional Fields

Add up to 5 fields to show on the cards.

Select Field

Select Field

Select Field

Select Field

Select Field

OK

Cancel

**FIGURE 19-9:** The Customize Task Board Cards dialog box.

*© John Wiley & Sons, Inc.*

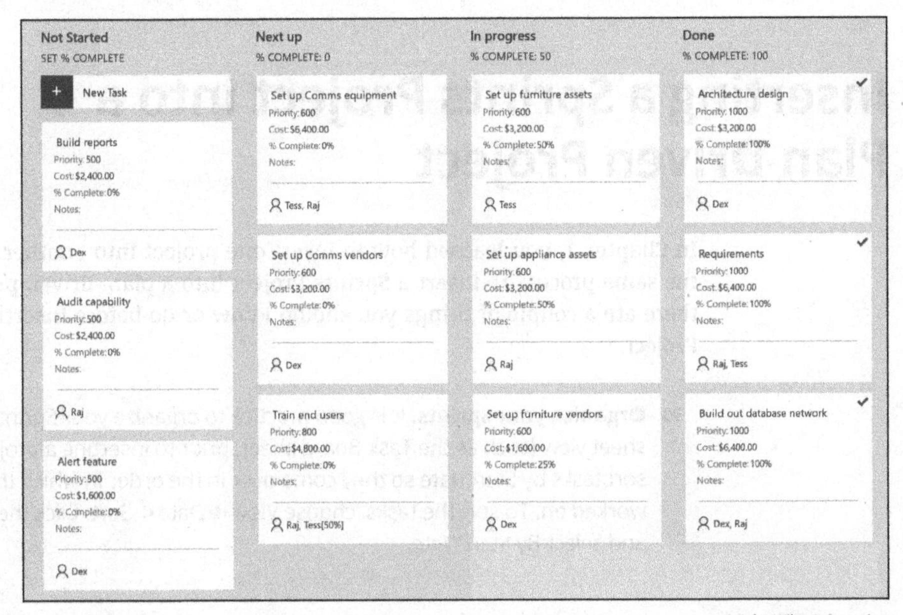


**FIGURE 19-10:** Customized cards.

*© John Wiley & Sons, Inc.*

## Prioritizing Tasks

One of the benefits of a Sprints Project is that work in the backlog can be reprioritized based on feedback, changes in stakeholder needs, or changes in the environment. Project has a field called Priority. It is on the General tab in the Task Information dialog box. This field is a great way to prioritize sprints work. To show the priority, just insert a column into a sheets view or add the Priority field to the Task Board cards.

Figure 19-11 shows the Priority column for the Asset Management project. You can see that tasks are ordered by priority. The project team will use this information to determine which tasks to work on in the next sprint. This approach ensures that the most important work is accomplished first.

| ⓘ | Sprint | Name | Work | Board Status | % Complete | Priority |
|---|---|---|---|---|---|---|
| ✓ | Set up | Set Up | 120 hrs | Done | 100% | 1000 |
| ✓ | Requirements | Architecture design | 40 hrs | Done | 100% | 1000 |
| ✓ | Requirements | Requirements | 80 hrs | Done | 100% | 1000 |
| ✓ | Database | Build out database network | 80 hrs | Done | 100% | 1000 |
| ✓ | Database | Set up users and permissions | 40 hrs | Done | 100% | 1000 |
| ✓ | Vehicles | Load vehicle assets | 20 hrs | Done | 100% | 800 |
| ✓ | Vehicles | Train end users | 60 hrs | Done | 100% | 800 |
| | Communication Assets | Set up furniture assets | 40 hrs | In progress | 50% | 600 |
| | Communication Assets | Set up appliance assets | 40 hrs | In progress | 50% | 600 |
| | Communication Assets | Set up furniture vendors | 20 hrs | In progress | 25% | 600 |
| | Communication Assets | Set up appliance vendors | 20 hrs | In progress | 25% | 600 |
| | Furniture Assets | Set up Comms equipment | 80 hrs | Next up | 0% | 600 |
| | Furniture Assets | Set up Comms vendors | 40 hrs | Next up | 0% | 600 |

**FIGURE 19-11:** Prioritizing tasks.



## Inserting a Sprints Project into a Plan-Driven Project

In Chapter 2, you learned how to insert one project into another. You can follow the same process to insert a Sprints Project into a plan-driven project. However, there are a couple of things you should know or do before inserting your Sprints Project.

- **Organize your sprints.** It is good practice to organize your Sprints Project in a sheet view (such as the Task Board sheet) prior to inserting a project. I like to sort tasks by Start date so they come over in the order in which they will be worked on. To sort the tasks, choose View ➪ Data ➪ Sort. Click the down arrow and select By Start Date.

- **Sprints and board status information won't be visible.** Because plan-driven projects don't have sprints, the only information that will be visible for the Sprints Project is the information shown in the Master Project. If you want to see information about sprints or board status, you can add those columns to your Gantt chart. All the tasks that are not part of the Sprints Project show No Sprint in the Sprints column and Not Started in the Board Status column.

Figure 19-12 shows a section of a plan-driven project with a Sprints Project inserted. Notice that the Sprint column and the Board Status column have been added.

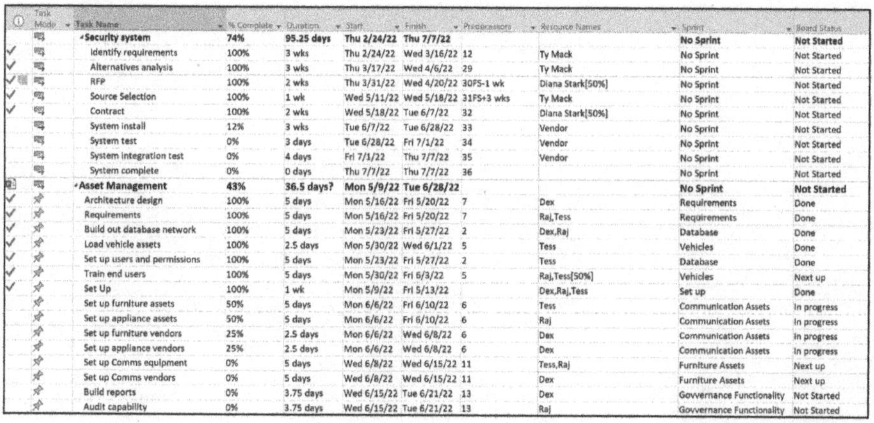

| ⓘ | Task Mode | Task Name | % Complete | Duration | Start | Finish | Predecessors | Resource Names | Sprint | Board Status |
|---|---|---|---|---|---|---|---|---|---|---|
| | | **Security system** | **74%** | **95.25 days** | **Thu 2/24/22** | **Thu 7/7/22** | | | **No Sprint** | **Not Started** |
| ✓ | | Identify requirements | 100% | 3 wks | Thu 2/24/22 | Wed 3/16/22 | 12 | Ty Mack | No Sprint | Not Started |
| ✓ | | Alternatives analysis | 100% | 3 wks | Thu 3/17/22 | Wed 4/6/22 | 29 | Ty Mack | No Sprint | Not Started |
| ✓ | | RFP | 100% | 2 wks | Thu 3/31/22 | Wed 4/20/22 | 30FS-1 wk | Diana Stark[50%] | No Sprint | Not Started |
| ✓ | | Source Selection | 100% | 1 wk | Wed 5/11/22 | Wed 5/18/22 | 31FS+3 wks | Ty Mack | No Sprint | Not Started |
| ✓ | | Contract | 100% | 2 wks | Wed 5/18/22 | Tue 6/7/22 | 32 | Diana Stark[50%] | No Sprint | Not Started |
| | | System install | 12% | 3 wks | Tue 6/7/22 | Tue 6/28/22 | 33 | Vendor | No Sprint | Not Started |
| | | System test | 0% | 3 days | Tue 6/28/22 | Fri 7/1/22 | 34 | Vendor | No Sprint | Not Started |
| | | System integration test | 0% | 4 days | Fri 7/1/22 | Thu 7/7/22 | 35 | Vendor | No Sprint | Not Started |
| | | System complete | 0% | 0 days | Thu 7/7/22 | Thu 7/7/22 | 36 | | No Sprint | Not Started |
| | | **Asset Management** | **43%** | **36.5 days?** | **Mon 5/9/22** | **Tue 6/28/22** | | | **No Sprint** | **Not Started** |
| ✓ | | Architecture design | 100% | 5 days | Mon 5/16/22 | Fri 5/20/22 | 7 | Dex | Requirements | Done |
| ✓ | | Requirements | 100% | 5 days | Mon 5/16/22 | Fri 5/20/22 | 7 | Raj,Tess | Requirements | Done |
| ✓ | | Build out database network | 100% | 5 days | Mon 5/23/22 | Fri 5/27/22 | 2 | Dex,Raj | Database | Done |
| ✓ | | Load vehicle assets | 100% | 2.5 days | Mon 5/30/22 | Wed 6/1/22 | 5 | Tess | Vehicles | Done |
| ✓ | | Set up users and permissions | 100% | 5 days | Mon 5/23/22 | Fri 5/27/22 | 2 | Tess | Database | Done |
| ✓ | | Train end users | 100% | 5 days | Mon 5/30/22 | Fri 6/3/22 | 5 | Raj,Tess[50%] | Vehicles | Next up |
| ✓ | | Set Up | 100% | 1 wk | Mon 5/9/22 | Fri 5/13/22 | | Dex,Raj,Tess | Set up | Done |
| | | Set up furniture assets | 50% | 5 days | Mon 6/6/22 | Fri 6/10/22 | 6 | Tess | Communication Assets | In progress |
| | | Set up appliance assets | 50% | 5 days | Mon 6/6/22 | Fri 6/10/22 | 6 | Raj | Communication Assets | In progress |
| | | Set up furniture vendors | 25% | 2.5 days | Mon 6/6/22 | Wed 6/8/22 | 6 | Dex | Communication Assets | In progress |
| | | Set up appliance vendors | 25% | 2.5 days | Mon 6/6/22 | Wed 6/8/22 | 6 | Dex | Communication Assets | In progress |
| | | Set up Comms equipment | 0% | 5 days | Wed 6/8/22 | Wed 6/15/22 | 11 | Tess,Raj | Furniture Assets | Next up |
| | | Set up Comms vendors | 0% | 5 days | Wed 6/8/22 | Wed 6/15/22 | 11 | Dex | Furniture Assets | Next up |
| | | Build reports | 0% | 3.75 days | Wed 6/15/22 | Tue 6/21/22 | 13 | Dex | Govvernance Functionality | Not Started |
| | | Audit capability | 0% | 3.75 days | Wed 6/15/22 | Tue 6/21/22 | 13 | Raj | Govvernance Functionality | Not Started |

**FIGURE 19-12:** A Sprints Project inserted into a plan-driven project.

© John Wiley & Sons, Inc.

As a reminder, use these steps to insert one project into another.

1. **In Gantt Chart view, select the task in the Task Name column above which you want the other project to be inserted.**
2. **From the Project tab, in the Insert group, select Subproject.**

   The Insert Project dialog box appears.
3. **Using the navigation pane and file list, locate the file that you want to insert and click it to select it.**
4. **If you want to link to the other file so that any updates to it are reflected in the copy of the project you're inserting, make sure that the Link to Project check box is selected.**
5. **Click the Insert button to insert the file.**

   The inserted project appears above the task you selected when you began the insert process. You can click the drop-down button to the right of the Insert button and then click Insert Read-Only if you just want people to be able to view the file, but not make any changes to it.

IN THIS CHAPTER

» Viewing your Sprints Projects with tables

» Filters and sorting Sprints Projects

» Creating sprint reports

# Chapter 20
# Tracking a Sprints Project

Working on Sprints Projects comes with a whole set of practices that are specific to them. You may have heard of sprint planning meetings, demos, retrospectives, daily scrum, and so forth. It's whole different mindset and comes with its own language. Describing that mindset is beyond the scope of this book, so if you want to learn more, you can check out *Agile Project Management For Dummies* by Mark Layton (Wiley). This book focuses on the mechanics of tracking Sprints Projects, not the philosophy and mindset of leading them.

This chapter shows you how to use tables, filters, and groups to focus on specific aspects of a project. You look at how to track a stand-alone Sprints Project as well as one that is integrated into a larger project. You also look at reports that are specifically designed for Sprints Projects. By the time you read through the chapter, you will be a virtuoso with Sprints Projects as well as plan-driven projects!

## Viewing Your Sprints Project Data

Leave it to Project to have special tables and filters for Sprints Projects. You can use tables, filters, and groups to arrange your data however you like! And you can do this all for a stand-alone Sprints Project or one that is merged with another project.

## Using filters to focus

You can use filters to show only the information you want to see. This is particularly helpful when you only want to review information for Sprints Projects.

REMEMBER

You access the Filter drop-down list in the Data group on the View tab.

There are several filters that are specific to Sprints Projects. Here are a few of them.

- **Backlog:** This filter shows tasks that are active in the backlog — in other words, they aren't done.
- **Current Sprint:** Shows tasks in the current sprint, regardless of their status.
- **Current Sprint Remaining Tasks:** Shows tasks in the current sprint that aren't done.
- **Select Sprint:** When you select this filter, a dialog box appears. You type the name of the sprint you want to see.

Figure 20-1 shows the Backlog filtered for remaining tasks.

| Name | Work | Board Status | Resource Names |
|---|---|---|---|
| Set up furniture assets | 40 hrs | In progress | Tess |
| Set up appliance assets | 40 hrs | In progress | Raj |
| Set up furniture vendors | 20 hrs | In progress | Dex |
| Set up appliance vendors | 20 hrs | In progress | Dex |
| Set up Comms equipment | 80 hrs | Next up | Tess,Raj |
| Set up Comms vendors | 40 hrs | Next up | Dex |
| Build reports | 30 hrs | Not Started | Dex |
| Audit capability | 30 hrs | Not Started | Raj |
| Alert feature | 20 hrs | Not Started | Dex |
| Integrate | 40 hrs | Not Started | Tess |

**FIGURE 20-1:** Filtered for remaining tasks.



## Using tables to arrange data

A table in Project is a preselected set of fields that you can use to view project data. You can read about Cost tables in Chapter 10 and Tracking tables in Chapter 15. This section covers four tables that are specific to Sprints Projects.

REMEMBER

To see the tables available to you, click the View tab and go to the Data group. Click the Tables button, and then choose More Tables.

The Sprint Planning and the Current Sprint tables have similar fields. Both tables have the following fields:

- Sprint
- Name
- Work
- Board Status
- Resource Names
- Task Summary Name
- Deadline

The Sprint Planning table sorts based the status of each task, such as Not Started, Next Up, and so on. This table also has a field for Show on Board. The Current Sprint table sorts based on sprint, so the tasks are grouped by sprint.

TIP

The Task Summary Name refers to the Summary Task if you are using an outline format, such as in a Gantt project. The Deadline field is more applicable to plan-driven rather than Sprints Projects. Since I don't usually use these fields for Sprints Projects, I hide those columns.

The two tables — Task Board Tasks and Backlog — have similar information as the sprint tables. The Task Board Tasks table sorts by sprint. The Backlog table doesn't have a field for sprint, so it sorts by status.

If this all seems a bit confusing, it is because all these tables basically show the same information, they just arrange it differently. The following table summarizes the content in each table.

| Table | Sprint | Name | Work | Board Status | Resource Names | Task Summary Name | Deadline | Show on Board |
|---|---|---|---|---|---|---|---|---|
| Current Sprint | ✓ | ✓ | ✓ | ✓ | ✓ | ✓ | ✓ | |
| Sprint Planning | ✓ | ✓ | ✓ | ✓ | ✓ | ✓ | ✓ | ✓ |
| Task Board Tasks | ✓ | ✓ | ✓ | ✓ | ✓ | ✓ | ✓ | |
| Backlog | | ✓ | ✓ | ✓ | ✓ | ✓ | ✓ | |

If you are looking at these tables when the Sprints Project is inserted into another project, the tasks that are not part of the Sprints Project will appear, but they all say No Sprint. If you just want to focus on the sprints tasks, you can apply the Backlog filter.

## Being a groupie

Another great way to organize and view your data is by grouping it. There are three groups that are specific for Sprints Projects: Board Status, Priority, and Sprint. Figure 20-2 shows a task grouped by priority.

| Name | Work | Board Status |
|---|---|---|
| **Priority: 200 - 299** | **0 hrs** | |
| Inventory management feature | 0 hrs | Not Started |
| Finance interface | 0 hrs | Not Started |
| Queries feature | 0 hrs | Not Started |
| Define compliance and governance | 0 hrs | Not Started |
| Build compliance features | 0 hrs | Not Started |
| Build governance features | 0 hrs | Not Started |
| **Priority: 300 - 399** | **0 hrs** | |
| Load buildings | 0 hrs | Not Started |
| Load systems assets and vendors | 0 hrs | Not Started |
| Load building vendors | 0 hrs | Not Started |
| Load building vendors | 0 hrs | Not Started |
| Integration | 0 hrs | Not Started |
| **Priority: 400 - 499** | **0 hrs** | |
| Set up grounds assets | 0 hrs | Not Started |
| Set up grounds vendors | 0 hrs | Not Started |
| Set up maintenance assets | 0 hrs | Not Started |
| Set up maintenance vendors | 0 hrs | Not Started |
| Set up materials and supplies | 0 hrs | Not Started |
| Set up material and supply vendors | 0 hrs | Not Started |
| Integration | 0 hrs | Not Started |

**FIGURE 20-2:** Tasks grouped by priority.



To group tasks, click the View tab and go to the Data group. Click the Group By drop-down button, and then choose how you want to group your tasks.

## Sorting tasks

Another option for viewing your tasks is by sorting. Click the View tab, go to the Data group, and then click Sort. You will see options to sort by start date, finish date, priority, cost, and ID. However, if you want more options, click Sort By to display the Sort dialog box. When you click the Sort By arrow, you will see a whole array of options for sorting. Some of the options useful for Sprints Projects include Board Status, Sprint, Sprint ID, Show on Board, and Priority.

There are pros and cons to sorting. One of the great things about sorting is you can sort by multiple variables, such as by sprint and then board status. You can also indicate whether you want to see the data in ascending or descending order.

Unfortunately, because the information isn't grouped, it can be confusing. Figure 20-3 shows how I sorted by sprint and board status, but you can't see what the sprint is unless you insert the Sprint column.

| Name | Work | Board Status |
|---|---|---|
| Load buildings | 0 hrs | Not Started |
| Load systems assets and vendors | 0 hrs | Not Started |
| Load building vendors | 0 hrs | Not Started |
| Load building vendors | 0 hrs | Not Started |
| Integration | 0 hrs | Not Started |
| Set up furniture assets | 40 hrs | In progress |
| Set up appliance assets | 40 hrs | In progress |
| Set up furniture vendors | 20 hrs | In progress |
| Set up appliance vendors | 20 hrs | In progress |
| Inventory management feature | 0 hrs | Not Started |
| Finance interface | 0 hrs | Not Started |
| Queries feature | 0 hrs | Not Started |
| Define compliance and governance | 0 hrs | Not Started |
| Build compliance features | 0 hrs | Not Started |
| Build governance features | 0 hrs | Not Started |
| Set up Comms equipment | 80 hrs | Next up |
| Set up Comms vendors | 40 hrs | Next up |

**FIGURE 20-3:** Tasks sorted by sprint and then board status.



# Creating Sprints Reports

Project has some truly great Task Board reports, and they are set up only to include work and tasks that are on Task Boards. This means you can focus on the Task Board and sprint work rather than get distracted by other project work. You can access these reports by clicking the Report tab, choosing the View Reports group, and then clicking Task Boards. You will see five predesigned reports:

» **Boards – Task Status:** Shows information on the task status for all the tasks in the project.

- **Boards – Work Status:** Shows information on the actual and remaining work for the project.
- **Current Sprint – Task Status:** Shows various views of the tasks in the current sprint.
- **Current Sprint – Work Status:** Shows various views of actual and remaining work.
- **Sprint Status:** Shows the number of tasks per sprint and the hours of work per sprint.

Just because these reports are predesigned doesn't mean you can't tailor them to meet your needs. By clicking one of the sections in the report, a Field List appears (as shown in Figure 20-4) that allows you to select which fields you want in a particular section of the report. You can also modify reports by clicking the Report Design contextual tab shown in Figure 20-5. From there, you can apply themes, change colors, insert shapes or images, and so on.

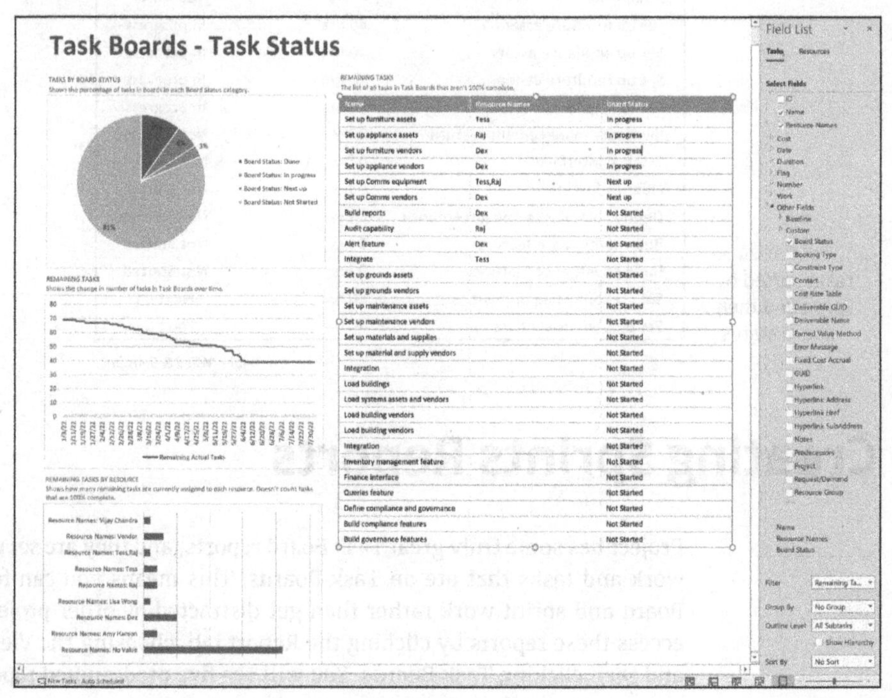


**FIGURE 20-4:** Task Status report.

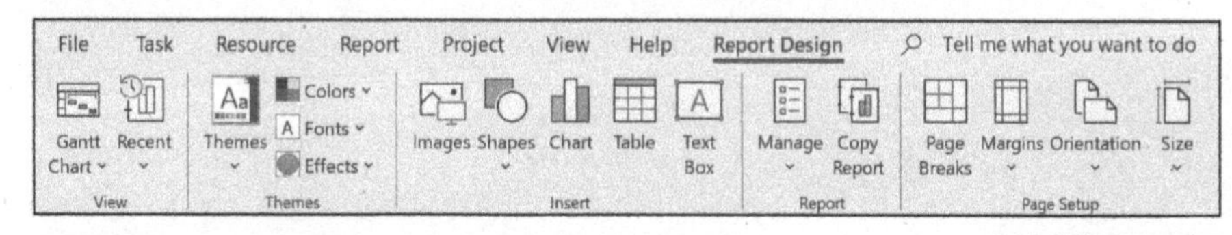


**FIGURE 20-5:** The Report Design contextual tab.

© John Wiley & Sons, Inc.

The reports for the Boards – Task Status and Work Status reports consider the overall project. The Current Sprint reports focus on the immediate work, as shown in Figure 20-6. These reports provide helpful information about actual and remaining work, a burndown chart for the project overall, work by resource, and remaining tasks.

REMEMBER

For more information on reports in general, you can review Chapter 18.

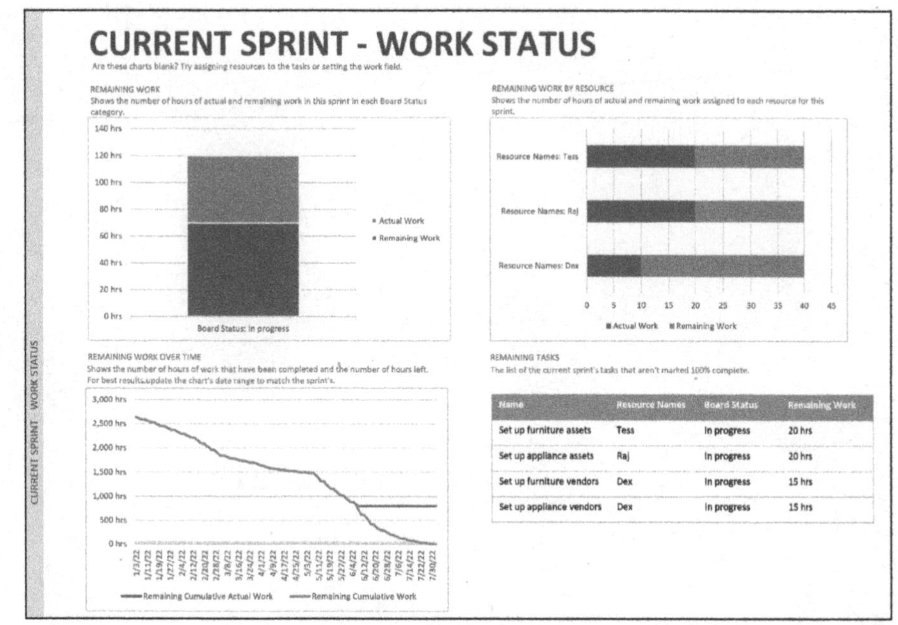


**FIGURE 20-6:** The Current Sprint report.

© John Wiley & Sons, Inc.

IN THIS CHAPTER

» Reviewing your successes and failures

» Comparing versions of the same project

» Creating a template for future projects

» Mastering the Organizer tool

# Chapter 21
# Getting Better All the Time

Have you ever finished a project and then wondered how in the heck it turned out the way it did? The budgeted totals mystically appear to be significantly over the original estimates and you missed the final deadline by three weeks! But you delivered the deliverables (somehow), and you can finally stuff the project file in the bottom of a drawer. Or can you?

Don't think of Microsoft Project as simply a giant electronic to-do list; rather, it's a sophisticated tool that's used to manage projects. And the logical byproduct of that management is a fantastic treasure-trove of information you can use to become a more highly skilled Project user — and, thus, a more competent project manager.

After you send the last memo related to the project and accept the final kudos or criticism from your boss, take a moment to look over the Project schedule one more time.

## Reviewing the Project

When I teach classes in project management, my students are often overwhelmed by all the information they have to collect, analyze, and report. Add to these elements a product such as Microsoft Project, and many times students begin reeling from all the data that they have to input and all the information that Project spews at them. They can't see straight for all the views, reports, tables, and filters they have access to in order to get information about their projects.

TIP

I tell them this secret: You won't understand every aspect of managing your first, second, or even third project. But you will become more comfortable with your skills. Similarly, you won't understand every nuance of Project on your first project. You won't even uncover every capability of Project on your second project. But gradually, as you master the ins and outs of managing projects and using Project, you'll apply project management skills more effectively and absorb the information that Project generates more easily — and you'll understand how that information can help you avoid mistakes on future projects.

Review every project at the end of each phase and at the end of the project to see what you did right and what you did wrong. It is also a good idea to reflect on your process if a major risk or issue occurs. Then use what you discover to improve your results during the rest of the project.

Sprints Projects often conduct a retrospective at the end of each sprint. A *retrospective* is a meeting where the team considers how to improve their performance and results in future sprints. A retrospective is an excellent tool for continuous improvement.

### Learning from your mistakes

You know what they say: If you don't study history, you're doomed to repeat it. And repeating a mistake is the last thing you want to do when managing a project.

Consider these strategies when reviewing your project:

- **Use a Variance table to compare the original baseline plan against the final actual activity.** Even if you created interim or baseline plans to adjust for drastic changes, look at the widest gap between what you expected to happen in the initial plan and what did happen. This strategy can be the best way to see areas where you tend to underestimate the most or where changes in the project's scope caused dramatic changes. Figure 21-1 shows a Variance table. For a refresher on baselines, refer to Chapter 14.

- **Review the notes you made about tasks to remind yourself of changes or problems that cropped up.** Insert the column named Notes into the Gantt Chart sheet and read all notes in one sitting.
- **Note which resources failed to deliver as promised; if you manage them, provide them constructive feedback.** If you don't manage resources, maintain notes on their working speed and quality to assist in future assignments. Also note which outside vendors failed to perform as expected and why this happened.
- **Assess your own communications in saved emails or memos.** Determine whether you gave the team enough information to perform effectively and whether you informed management about project issues in a timely way.
- **Assess feedback from your boss, your customers, and other stakeholders.** Determine whether any overall constraints, such as the project budget or deliverables, changed dramatically during the course of the project. Ask whether end users were satisfied with the rate at which you completed the project and delivered the final product.
- **Meet with your team.** Ask your team what worked well, what can be improved, and whether they would rely on you to lead another, similar project.

| Task Name | Start | Finish | Baseline Start | Baseline Finish | Start Var. | Finish Var. |
|---|---|---|---|---|---|---|
| **Perimeter** | **Mon 1/3/22** | **Fri 4/8/22** | **Mon 1/3/22** | **Fri 4/8/22** | **-0.88 days** | **0 days** |
| **Walls** | **Mon 1/3/22** | **Fri 3/11/22** | **Mon 1/3/22** | **Fri 3/11/22** | **-0.88 days** | **-0.75 days** |
| Dig trench | Mon 1/3/22 | Fri 1/14/22 | Mon 1/3/22 | Mon 1/17/22 | -0.88 days | -0.88 days |
| Install footing | Mon 1/17/22 | Fri 1/21/22 | Mon 1/17/22 | Tue 1/25/22 | -0.88 days | -2 days |
| Install cinder blocks | Mon 1/24/22 | Fri 2/11/22 | Mon 1/24/22 | Wed 2/16/22 | -0.88 days | -3 days |
| Apply stucco | Mon 2/14/22 | Fri 2/18/22 | Mon 2/14/22 | Mon 2/21/22 | -0.88 days | -0.88 days |
| Paint | Mon 2/21/22 | Mon 2/28/22 | Tue 2/22/22 | Mon 2/28/22 | -1 day | -0.75 days |
| Apply finishes | Mon 2/28/22 | Fri 3/4/22 | Mon 2/28/22 | Fri 3/4/22 | -0.75 days | -0.75 days |
| Wall contingency | Fri 3/4/22 | Fri 3/11/22 | Mon 3/7/22 | Fri 3/11/22 | -0.75 days | -0.75 days |
| Walls complete | Fri 3/11/22 | Fri 3/11/22 | Fri 3/11/22 | Fri 3/11/22 | -0.75 days | -0.75 days |
| **Entry gates** | **Tue 2/15/22** | **Fri 4/8/22** | **Mon 2/14/22** | **Fri 4/8/22** | **0.13 days** | **0 days** |
| Select gates | Tue 2/15/22 | Thu 2/24/22 | Mon 2/14/22 | Mon 2/28/22 | 0.13 days | -1.88 days |
| Purchase gates | Mon 2/28/22 | Tue 3/29/22 | Mon 2/28/22 | Mon 3/21/22 | -0.88 days | 5.38 days |
| Install gates | Mon 4/4/22 | Wed 4/6/22 | Wed 3/30/22 | Fri 4/1/22 | 3 days | 3 days |
| Install card readers | Mon 4/4/22 | Mon 4/4/22 | Mon 4/4/22 | Mon 4/4/22 | 0 days | 0 days |
| Install infrared photo devices | Fri 4/8/22 | Fri 4/8/22 | Fri 4/8/22 | Fri 4/8/22 | 0 days | 0 days |
| Perimeter Complete | Fri 4/8/22 | Fri 4/8/22 | Fri 4/8/22 | Fri 4/8/22 | 0 days | 0 days |

**FIGURE 21-1:** The Variance table.



## Fine-tuning communication

No project is the province of a single person. Even if no one else ever touched the Project plan, your team still provided input to that plan via the hours of activity they reported and the information they provided to you during the course of the project.

Follow these suggestions to refine the communications process:

- **Ask team members about the success of actual activity reporting.** Did you send email, schedule notes, or use the Project Web App tools, such as the Timesheet, to gather resource information? Did you consider taking advantage of all the benefits of online collaboration? Should you do so on the next project?
- **Determine whether your team rates your communications as frequent and thorough.** Did you share with resources too little information about the project, or did you inundate them with too much? Did you send simple reports on specific aspects of the project or send entire (cumbersome) Project files? Did the management team feel that your reporting on the project was sufficient for their needs? Should you learn to take better advantage of other software, such as Excel and Visio, accessed via Visual Reports?

# Comparing Versions of a Project

Project managers sometimes prefer to save new versions of project plans rather than work with interim plans. This approach can work, but it doesn't enable the project history to be captured in a single project file. No matter — Project's Compare Versions feature lets you see the differences between two Project plan files. These files can be either two different versions of the plan for a single project or files for two separate projects that happen to be similar.

You can use this method to review events from a completed project or to compare similar new projects as you are building them. The finished comparison report shows you the project differences graphically and provides a legend to identify report features.

To compare two files in Project, follow these steps:

1. **Open the file that you want to compare.**

   The active file is considered the current project.

2. **From the Report tab, click the Compare Projects button in the Project group.**

   The Compare Project Versions dialog box appears, as shown in Figure 21-2.

3. **If the file to compare doesn't appear in the drop-down list labeled Compare the Current Project *(filename)* to This Previous Version, click the Browse button to search for and select the appropriate file.**

   The file you select is considered the previous project.

4. **To compare tables other than the default tables that are shown, select the appropriate tables from the Task Table drop-down list and Resource Table drop-down list.**

5. **Click the OK button.**

6. **If you see a message indicating that you have too many columns to compare, click OK to continue.**

   Project runs the comparison and displays the comparison report, as shown in Figure 21-3. The comparison report appears in the upper pane, and the individual project files show up in the two lower panes. The legend at the left describes the symbols in the Indicators column of the report.

7. **Click Resource Comparison in the View group on the Compare Projects tab.**

   The Comparison Report pane changes to display resource information.

8. **Click the Close Comparison (X) button on the Legend for Comparison Report pane to close the pane.**

   Click the Save button on the Quick Access toolbar and name the file so that the next time you want to view this comparison report, the projects and fields will already be set up.

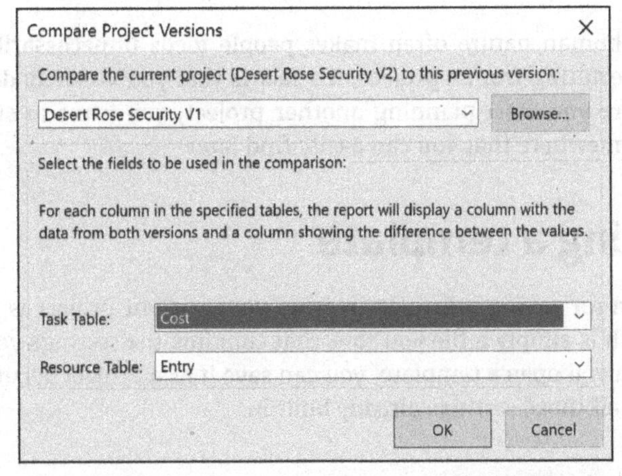


**FIGURE 21-2:** The Compare Project Versions dialog box.

© John Wiley & Sons, Inc.

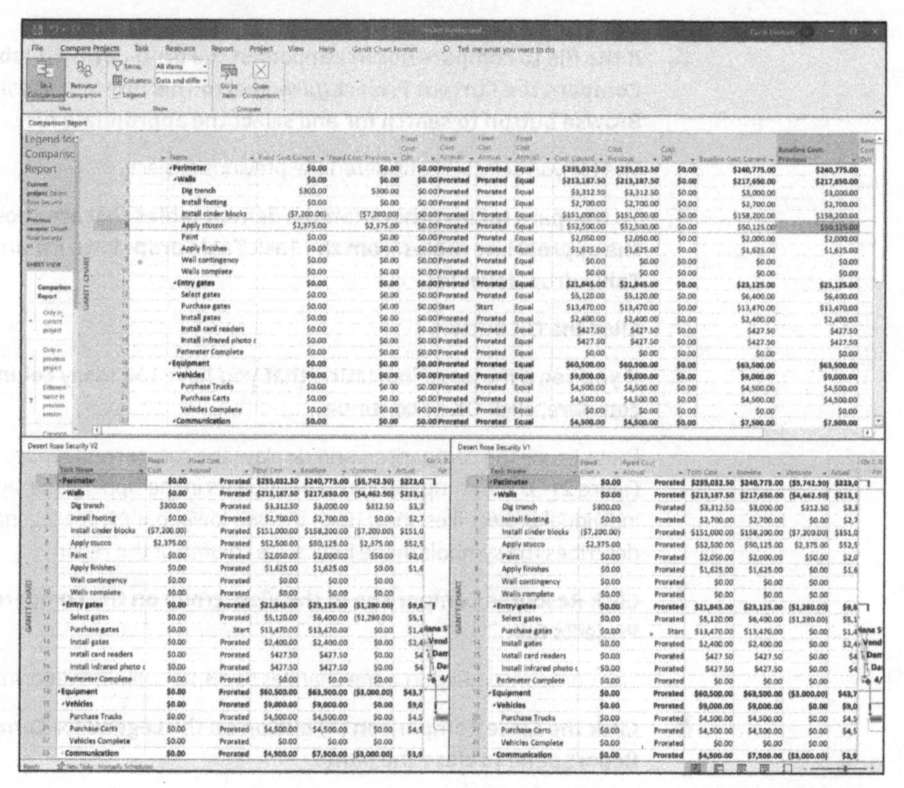

**FIGURE 21-3:** Comparing projects.

© John Wiley & Sons, Inc.

# Building on Success

Although human nature often makes people focus unnecessarily on the shortcomings resulting from a project, the fact is that you've likely done *many* things well. Before you start planning another project, you want to store the positive aspects somewhere that you can easily find later.

## Creating a template

One option for saving information from your current project is to create a *template*, which is simply a file you save that contains the settings you want to replicate. When you open a template, you can save it as a Project document with a new name and all those settings already built in.

Project contains its own templates for common types of projects, but you can save any project as a template. If you often repeat the same set of tasks in a project — as people in many industries do — save yourself the time of re-creating all those tasks.

Templates can contain, in addition to any tasks in the project, any or all of these types of information for those tasks:

- All information for each baseline
- Actual values
- Rates for all resources
- Fixed costs

You can save all this information or only selected items. For example, if you create numerous fixed costs (such as equipment) and resources with associated rates (and you use them in most projects), you can save a template with only fixed costs and resource rates.

To save a file as a template, follow these steps:

1. **Open the file you want to save and then choose File ⇨ Save As.**
2. **In the Save As column, select where you want to save the template.**

   You can save it in SharePoint, on your computer, in the cloud, or in another location that you add.
3. **Navigate to where you want to save the template.**

   The Save As dialog box appears, as shown in Figure 21-4.
4. **Name the template.**

   In the File Name field, type a name for your template.
5. **From the Save As Type drop-down list, select Project Template (*.mpt).**

   Be aware that this file extension doesn't appear unless you've set Windows to display file extensions.
6. **Click the Save button.**

   A dialog box appears (see Figure 21-5) that asks you to select the type of data you want to remove from the template, such as values of all baseline, actual values, and so on. Check the boxes to remove the applicable data from the template and click Save.

   Microsoft saves templates in the central folder named **Templates**.

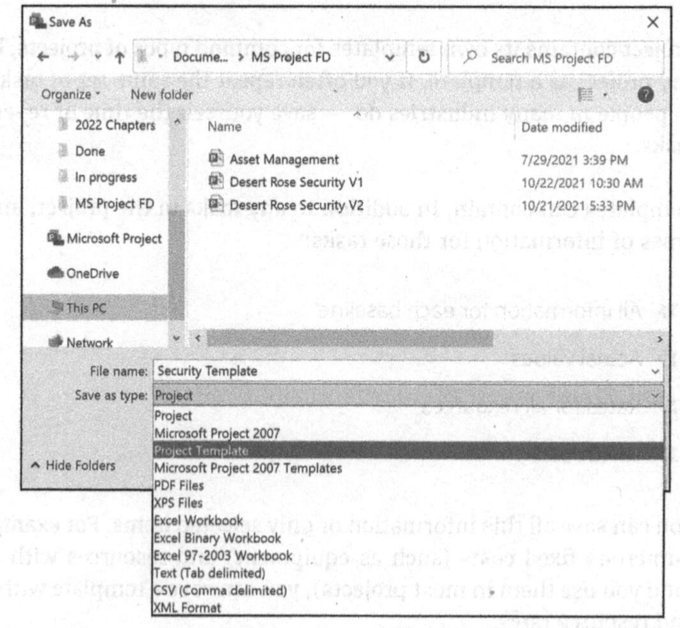


**FIGURE 21-4:** Saving a file as a template.

© John Wiley & Sons, Inc.

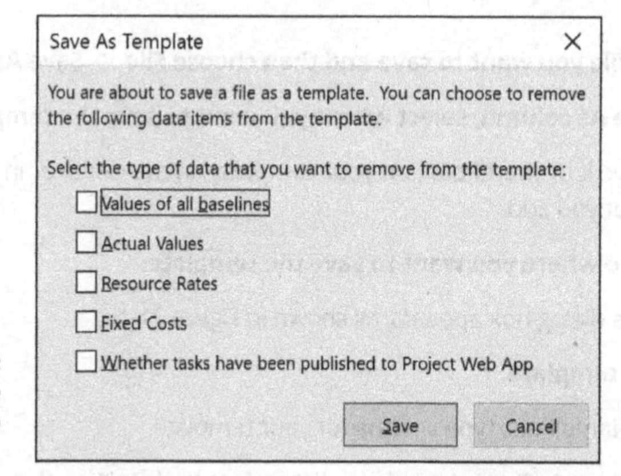


**FIGURE 21-5:** The Save As Template dialog box.

© John Wiley & Sons, Inc.

# Mastering the Organizer

The marvelously flexible nature of Project lets you customize many settings such as tables, filters, views, and reports. For example, you can create separate tables of data to display in views that contain sheet panes. You can also create separate filters, reports, and calendars. If you have any kind of a life, you may not want to spend evenings re-creating all these elements on the next project. Instead, use the Organizer to copy them to other Project files.

You can use the Organizer to copy information from one file to another file. There are nine tabs to help organize the information you can copy:

- Views
- Reports
- Modules
- Tables
- Filters
- Calendars
- Maps
- Fields
- Groups

TIP

If you want a custom item that you created in the current file available for use in all Project plan files, use the Organizer to copy the item into the global template named `Global.mpt`. Though some items are copied to the Organizer by default, it never hurts to ensure that crucial items are saved there.

Follow these steps to use the Organizer:

1. **Open the project that you want to copy items from and the project to copy items to.**
2. **In the file you want to copy to, choose File ⇨ Info.**

   Backstage view opens.
3. **Click the Organizer button.**

   The Organizer dialog box appears, as shown in Figure 21-6.
4. **Click the tab for the type of information you want to copy.**
5. **In the left column, click the line to copy from.**

   By default, Project shows the `Global.mpt` template in the list on the left and the file you opened in the Organizer in the list on the right. If you want to copy items to another open file instead, choose that file by clicking the drop-down list below the file list on the right. (Depending on the tab you display, the drop-down list may be named Views Available In, Tables Available In, Reports Available In, and so forth.)
6. **Click the Copy button.**

   The item appears in the list on the right.

7. **(Optional) To rename a custom item that you've created,**

   (a) *Click it in the list on the right and then click the Rename button.*

   (b) *Enter a new name in the Rename dialog box that appears and click the OK button.*

8. **To copy additional items on the same tab, repeat Steps 5 and 6.**

9. **To copy an item on a different tab, repeat Steps 4–7.**

10. **When you finish copying items from one file to another, click the Close (X) button.**

TIP

If you're copying items to or from the global template, you don't need to open it explicitly.

WARNING

Don't rename default items in the Organizer. Renaming them can cause errors and malfunctions, such as the inability to display a particular view.

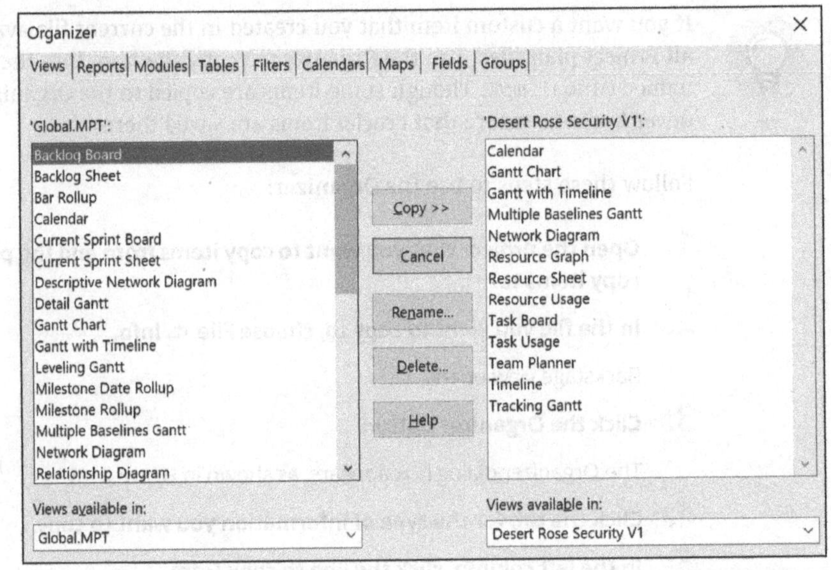


**FIGURE 21-6:** The Organizer dialog box.

# 6 The Part of Tens

IN THIS PART . . .

Check out the ten golden rules of project management.

Learn ten cool shortcuts in Project.

IN THIS CHAPTER

» Following good project-management practices

» Scheduling projects efficiently

» Communicating with the team

» Benefitting from mistakes

# Chapter 22

# Ten Golden Rules of Project Management

As you begin to use Project, the common sayings (or aphorisms, axioms, and precepts) in this chapter can help you recall basic project-management principles. Tack them on your office wall so that you can review them throughout the workday.

## Roll with It

Rolling wave planning is an excellent way to simplify the management of a project. Plan no further into the future than you can reasonably see — and don't plan in more detail than makes sense. When a project is first chartered, you only have milestone dates. As you begin to understand more about the project, you can define the project life cycle and the key deliverables to be created in each of its phases.

Build the project in phases. Start with the initial phases, and fill in detailed tasks, sequences, resources, and durations. Leave the later phases at a high level of detail; that way, you'll have less work to redo on later tasks that simply can't be anticipated at the beginning of the project. You'll also have less need to manipulate the baseline of those later tasks: If tasks are too far in the future, their time requirements won't match up with the resources you originally set aside for them.

Sprints Projects keep all the known work in a backlog. The backlog is prioritized (and reprioritized) based on evolving needs. Therefore the work in the current sprint is estimated and planned in detail. Work for future sprints is held at a higher level and is elaborated once that sprint begins.

REMEMBER

When it's early in the game, avoid delving into detail on tasks that are far in the future.

Project can help you with rolling wave planning when you use these features of the program:

- **Network Diagram view:** Helps you visualize project phases graphically
- **Timeline:** Shows milestones, deliverables, and phases at a high level
- **Task expansion and collapse:** Lets you hide or display project phases
- **Sprint planning:** Shows only the tasks scheduled for the current sprint

REMEMBER

See Chapter 3 for more about the best ways to collapse and expand summary tasks and subtasks.

## Put Your Ducks in a Row

Before you start creating the project, do your homework. If you don't have all the information you need when you sit down at the computer to work with Project, you'll continually stop midplan and run off to find the information — not an efficient way to work.

Before you sit down to build a project schedule, think about the following project information:

- **Reality check of basic expectations:** Based on your experience in managing similar projects over time, determine whether the deliverables, budget, and preliminary schedule are within reason. Decide whether to discuss changes to the scope and budget to avoid a project that's unrealistic from the start.
- **Resource information:** For team resources, record the resource's full name, contact information, skills, cost, schedule, timing conflicts, and manager's name and contact information. For equipment or facilities, find their availability and cost.

- **Communication guidelines:** Determine whether one team member records everyone else's progress or all resources do their own tracking. Decide who updates the schedule after changes are made, who receives copies of which reports, and who has access to the master schedule online.
- **Management expectations:** Ask whether management expects to see regularly scheduled basic reports or another type of report instead. Ask how, and from whom, to get budget approval at various phases in the project planning, and ask whether any cross-enterprise interests will require reports or approval from multiple sources. (See Chapter 18 for the lowdown on reports.)
- **Company policies:** These documents may spell out working hours and overtime policies, holiday calendars, charges for overhead costs or project markups, and the sharing (or not) of information with clients and vendors.

After you contemplate the issues in this list, you're ready to sit down and start entering information into Project.

# Expect the Unexpected

You know he's out there — Murphy and his darn law stipulating that anything that can go wrong *will* go wrong. Most projects, especially lengthier and more complex projects, aren't completed on time or on budget. Your job as the project manager is to plan as accurately as you can and then make prudent adjustments whenever someone throws a wrench into the works — and Project gives you lots of tools to do it. But beyond all the automated features of Project, you can anticipate change by simply planning for it.

The critical path determines the project duration. Wise project managers build contingency reserve time and funds into their projects. When a project wraps up a week late and $5,000 over budget, only the project manager knows that it was *four* weeks later and $25,000 costlier than originally scheduled, with no contingency reserve. (I introduce contingency reserve in Chapter 12.)

REMEMBER

Add contingency reserve before major deliverables or the end of a phase to account for unexpected events. You can even add a Contingency task and resource it with Murphy.

Use the cost resource type (rather than a work or material resource) to add a set amount of contingency reserve to a task or phase.

# Don't Put Off until Tomorrow . . .

Though Project management software can simplify many aspects of your work life, most people using Project for the first time become overwhelmed by the amount of time they spend entering and updating data. These tasks can certainly be cumbersome, but the reward from mastering the automated updating and reporting capabilities in Project more than makes up for any labor invested up front.

TIP

Where appropriate, import tasks from Outlook or Excel into Project to help speed data entry in the planning phase.

Track as often as you can — at least once a week. If you don't tend to the task of tracking progress on a project, you may wind up behind the proverbial eight ball. This strategy not only saves you from having to enter a mountain of tracking data, but also lets you — and your team — see the status of the project at any time. That way, you can promptly spot disaster approaching and make preventive adjustments.

# Delegate, Delegate, Delegate

Avoid the urge to attempt to do everything on a project yourself. Although creating and maintaining the Project file on your own might seem to give you more control over the result, flying solo in a larger project is nearly impossible. Of course, you can't allow dozens of people to make changes to the plan, because you would risk losing track of who did what and when. However, following these few simple practices can convert a few fingers in the project pie from harmful to helpful:

- **Designate one person to handle all data.** This person's mission is to enter all tracking data into the master file.
- **Break the project into a few subprojects and assign people whom you trust to act as managers of those subprojects.** Let them handle their own tracking and adjustments, and then assemble these projects into a master project so that you can monitor overall status.
- **Set uniform procedures for the team up front.** Don't let one team member report time on an interoffice memo sheet, another send you their progress by email, and others record their work hours in the timesheet willy-nilly. Work out a strategy that uses Project features to communicate consistently.

» **Share project management duties.** Sharing the responsibility for note taking, facilitating meetings, and giving presentations can free up some of your time and help your team members develop leadership skills. Sharing leadership also increases buy-in and commitment from the team.

REMEMBER

You can track progress at a high level, such as 25 percent, 50 percent, 75 percent, and 100 percent complete. (I discuss tracking in Chapter 15.)

# Document It

Most people have heard the project management warning "Cover your assumptions," and Project helps you do it easily. Use these features to document the details of the project:

» **Notes area:** Use it for both tasks and resources to make a record of background information, changes, or special issues.

» **Reports:** Customize them to incorporate all pertinent information and help document trends and changes, as detailed in Chapter 18.

» **Visual Reports:** Use this feature to paint the picture of the project status for visually oriented stakeholders; see Chapter 18.

Save multiple versions of the project, especially if you change the baseline in later versions. This way, you have a record of every step in the project planning process to refer to when questions arise down the road.

# Keep the Team in the Loop

I've worked in offices where I spent more time wrestling with the question of whom to keep informed about what than I did working. If I didn't include marketing and finance in every email on a new product launch, I'd be called on the carpet the next day, or (worse) a vital action step would fall through the cracks because someone didn't know to take action. Follow these methods of keeping communication channels open:

» **Use Project features that integrate with Outlook or other email programs.** Use them to send Project files or other types of communications.

- **Review progress with team members by meeting regularly in person, over the phone, or via meeting software.** Ensure that all team members have the latest version of the Project schedule to refer to during these meetings.
- **Display the work breakdown structure (WBS) code on reports.** Then you can easily refer to specific tasks in large projects without confusion. Work breakdown structures are covered Chapter 2.

# Measure Success

When you begin the project, you should have an idea of what constitutes success, and you should know how to measure that success. You've heard this one: "You can have the deliverable on time, on budget, or done right. Choose two." Success can involve attaining many goals, such as:

- Customer satisfaction
- Management satisfaction
- Meeting budget constraints
- Being on time

To determine how you'll measure success, ask these questions:

- Does success in budgeting mean not exceeding the original estimates by more than 10 percent?
- Is timely delivery based on meeting the number of working hours or meeting a specific deadline?
- How will you measure customer satisfaction?
- Does a successful product launch include high sales figures after the launch, or was the project successful merely because you moved it out the door?

TIP

Place milestones in the project that reflect the achievement of each type of success. When you reach one, you can pat your team on the back. Knowing what success looks like helps you motivate your team to get there.

## Maintain a Flexible Strategy

Stuff happens. There's never been a project that didn't require accommodations for surprises along the way. The mark of a good project manager is that they are alert to these changes and make adjustments to deal with them quickly.

Making adjustments to accommodate bad news isn't easy; in fact, it can be truly difficult to deliver bad news. However, avoiding a problem in the project and hoping that it will simply disappear has a nasty habit of snowballing into an even worse problem. The following tools can help you stay alert to changes and make adjustments:

- The work contour option can help you change the contour of the work to front-load, back-load, or otherwise change the spread of work over the duration of the task, as described in Chapter 9.
- Use the Task Inspector, detailed in Chapter 17, to see what's driving a task. You can look for options such as adding resources, revising task dependency relationships, and using reserve.
- Turn on critical path formatting to see the critical path of the project and track how much slack remains. Adjusting tasks to efficiently use up their slack can keep you on schedule in a crisis, as I describe in Chapter 17.

## Learn from Your Mistakes

One great gift that Project offers is the capability to look back after completing a project so that you can learn from your mistakes. You can review the original schedule, and every version after it, to see how well you estimated time and money, and then figure out how to do it better.

By using records of the project, you can spot trends to find out, for example, where you always seem to miss on timing, or why you always allow too little time for market research and too much time for questions and answers. Perhaps you always forget to budget for temporary help during rush periods, or you overstaff early on, when you need only a few people.

Use the wealth of information in Project schedules to educate yourself on your own strengths and weaknesses as a project planner and manager and to improve your skills with each project you take on.

IN THIS CHAPTER

» Viewing task and resource information

» Keeping shortcuts close at hand

» Changing the view for better focus

» Adding tasks to the Timeline

# Chapter 23 Ten Cool Shortcuts in Project

I'm always amazed to watch people work who are adept at using a specific type of software. Their fingers seem to fly over the keyboard; without ever touching the mouse, they create documents, presentations, and other artifacts. In this chapter, I show you ten cool shortcuts that can help you handle Project like a pro.

## Task Information

From any task view (including Task Boards view), you can double-click any task in the project to open the Task Information dialog box. You can then use it, as shown in Figure 23-1, to enter or modify durations, predecessor information, resources, notes, task types, and constraints.

In Chapter 2, I give you more information about the Task Information dialog box.

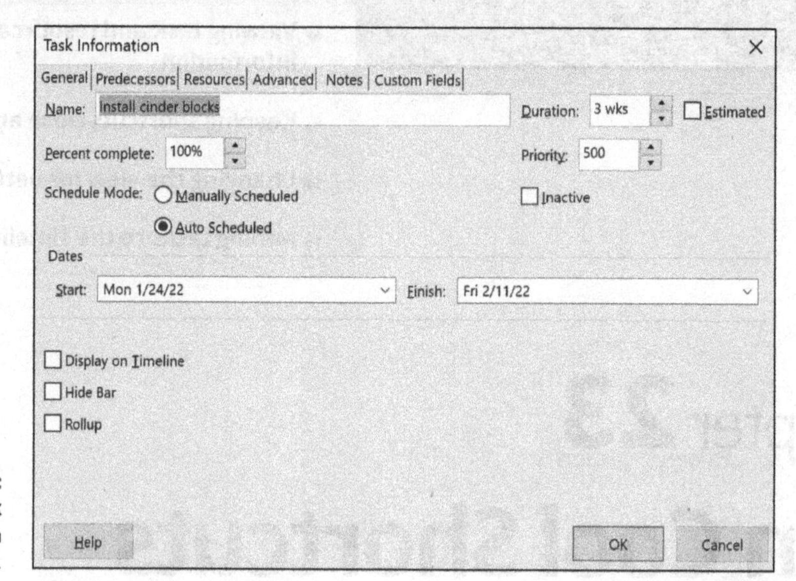


**FIGURE 23-1:** The Task Information dialog box.

© John Wiley & Sons, Inc.

## Resource Information

If you want to know anything about a resource, you can find the information in the Resource Information dialog box, as shown in Figure 23-2. From any Resource Sheet view, double-click a resource name to see contact information, availability, cost rates, and any notes you've entered about the resource.

You can use the Resource Information dialog box to enter, edit, or update resource information for people, supplies, equipment, and locations considered to be resources. In Chapter 7, I detail how to work with resources in Project.

TIP

When the Assign Resources dialog box is open, double-click a resource name to display the Resource Information dialog box.

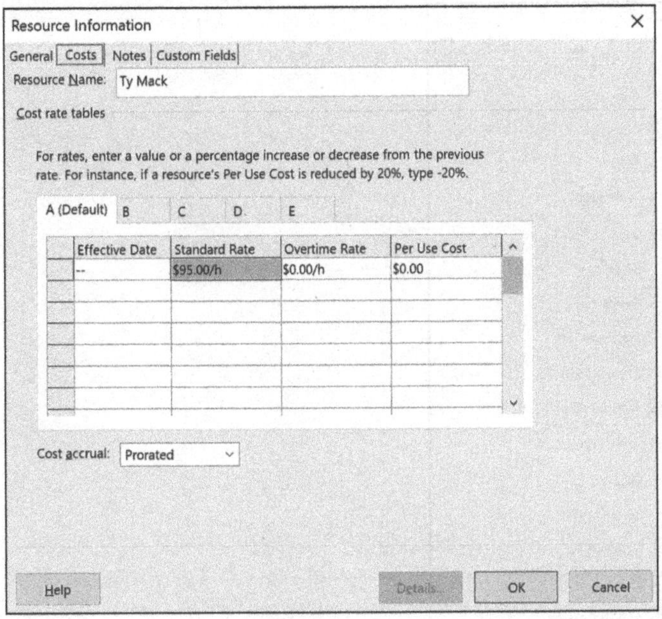

**FIGURE 23-2:** The Resource Information dialog box.

© John Wiley & Sons, Inc.

# Frequently Used Functions

Most user actions in Project are fairly standard, so you shouldn't have to visit the Ribbon every time you want to give a command. Instead, you can simply right-click to display a contextual menu of command options that let you (for example):

- Cut, copy, and paste
- Insert and delete tasks
- Assign resources
- Open the Task Information dialog box
- Add notes
- Add the task to the timeline

Figure 23-3 shows a typical contextual menu you see after right-clicking a task.

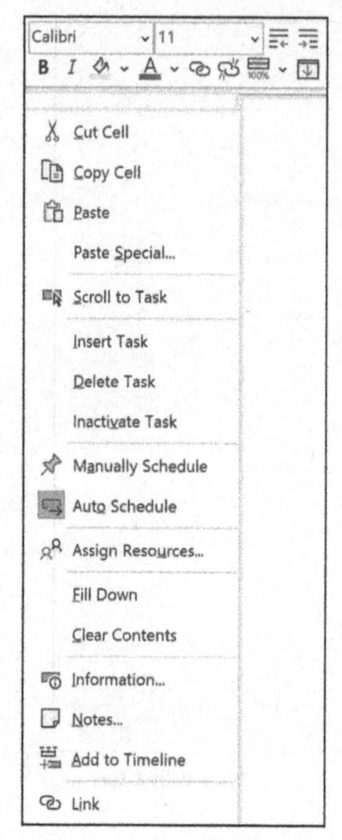


**FIGURE 23-3:** Contextual menu.

© John Wiley & Sons, Inc.

# Subtasks

Sometimes, you need to see portions of the project expanded to show all tasks in detail while other portions remain rolled up to a summary task level. In addition to the standard method of clicking the triangle next to the summary task to expand or summarize tasks, you can use keystrokes.

To hide a subtask, press Alt+Shift+hyphen (–). To show subtasks, press Alt+ Shift+plus sign (+).

# Quick Selections

Say you want to move a row. Place your cursor in any field on the row. You can quickly select the whole row by pressing Shift+Spacebar. To select a whole column, place your cursor anywhere on that column and press Ctrl+Spacebar.

## Fill Down

Suppose that all the tasks in a series have the same resource or the same duration. Rather than repeatedly enter the same resource or duration, you can enter the duration on the topmost task in the list, select the rest of the tasks, and then press Ctrl+D. Project "fills down" the resource to the rest of the selected tasks.

## Navigation

The Ctrl key is a helpful tool for navigating to the beginning or end of the project. Simply press Ctrl+Home to move to the beginning of the project or press Ctrl+End to move to the end.

You can also press the Ctrl+Alt keys to navigate around the timescale. To move to the right (forward in time), press Ctrl+Alt+right arrow. To move to the left (backward in time), press Ctrl+Alt+left arrow.

## Hours to Years

Sometimes, you want to see the big-picture view of the project by viewing the timescale at the level of months and years. At other times, you need to see details about when tasks will occur. In Project, you can set the timescale to show time in years, all the way down to hours. To show incrementally larger timescale units, press Ctrl+Shift+* (asterisk). To show smaller, more detailed timescale units, press Ctrl+/ (slash).

## Timeline Shortcuts

Even the Timeline has shortcuts. When you first show the Timeline, it displays the text `Add tasks with dates to the timeline`. Double-click the text to open the Add Tasks to Timeline dialog box (shown in Figure 23-4). Every task is accompanied by a check box; simply select the ones you want to show on the Timeline.

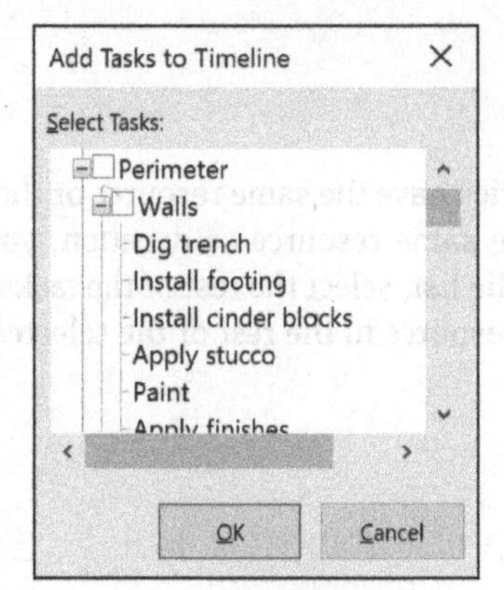


**FIGURE 23-4:** Adding tasks to the Timeline.

© John Wiley & Sons, Inc.

WARNING

**Double-clicking the initial text to open the Add Tasks to Timeline dialog box only works the first time you open the Timeline. After you begin entering information on the Timeline, the text that appears on it the first time you open it disappears, and you have to enter tasks by right-clicking them first and choosing Add to Timeline from the shortcut menu.**

# Quick Undo

**What if you're typing along and you find the last two rows were entered wrong. Easy! Press Ctrl+Z to undo your last entry. You can press it again to undo the one before that, and the one before that . . . you get the idea.**

# Glossary

**activity:** Also known as *task*, an individual step performed to reach a project's goal.

**actual:** Final reporting data for a task, such as the date when the task really started versus its planned start date or the amount of work completed versus the planned amount of work.

**ACWP (actual cost of work performed):** Cost of the actual, real work done on a task, work package, or project to date.

**Agile:** A mindset of values and principles set forth in the Agile Manifesto.

**ALAP (as late as possible):** A constraint put on a task's timing to make the task occur as late as possible in the project schedule, taking into account any dependency relationships.

**analogous estimate:** An estimating method that uses information from past similar projects and tailors it for the current project estimates.

**ASAP (as soon as possible):** A constraint put on a task's timing to make the task occur as early as possible in the project schedule, taking into account any dependency relationships. This is the default constraint type.

**auto-scheduled:** A task mode where Project calculates the schedule as needed based on task dependencies, constraints, calendars, and resource availability.

**BAC (budget at completion):** The sum total of all costs involved in completing a task, work package, or project. *See also* baseline cost.

**backlog:** A list of work that has to be done.

**Backstage view:** The area that presents files and options when you click the File tab on the Ribbon.

**baseline:** The formally agreed-upon dates for start, finish, duration, work, and cost data in the project. Later, actual schedule and duration, work, and cost data are tracked so they can be compared with the baseline data.

**baseline cost:** The formally agreed-upon planned costs for a project's tasks, used to compare with actual costs as they are incurred.

**BCWP (budgeted cost of work performed):** Also called *earned value;* the value of work that has been completed on a task, work package, or project. For example, a task with $1,000 of costs accrues a BCWP of $750 when it's 75 percent complete.

**BCWS (budgeted cost of work scheduled):** The budget for the scheduled task, work package, or project.

**booking type:** A category for resources that specifies whether they're committed to the project or simply proposed to be involved.

**burndown chart:** A report that compares the amount of work (or tasks) remaining to the baseline work (or tasks) remaining.

**calendar:** The various settings for hours in a workday, days in a workweek, holidays, and nonworking days on which a project schedule is based. You can set Project, Task, and Resource calendars.

**card:** A way of portraying task data on a Task Board. An electronic index card with data about a task.

**change highlighting:** A feature that highlights duration, start, and finish data changes caused by any other changes you've made in your project since you last saved it.

**charter:** A document that formally authorizes or recognizes a project.

**circular dependency:** A timing relationship among tasks that creates an endless loop that can't be resolved.

**collapse:** To close a project outline to hide subtasks from view.

**Combination view:** A Project view with task/resources details appearing at the bottom of the screen.

**constraint:** A parameter that forces a task to fit a specific timing. For example, a task can be constrained to start as late as possible in a project. Constraints interact with dependency links to determine a task's timing.

**contingency reserve:** Time or money that's factored into your schedule or budget to mitigate identified risks.

**cost:** The amount of money associated with a task when you assign *resources,* which are equipment, materials, or people with associated fees or hourly rates, and any additional fixed costs.

**cost variance (CV):** The difference between the baseline costs and the combination of actual costs to date and estimated costs remaining *(scheduled costs).* The cost variance is either *positive* (over budget) or *negative* (under budget). When using earned value management, CV is the difference between the earned value and the actual cost, shown as CV = EV–AC.

**critical path:** The series of tasks that must occur on time for the overall project to meet its deadline. The end date of the critical path is the earliest a project can be completed. Any slip on the critical path causes a slip in the due date.

**critical task:** A task on the critical path. *See also* critical path.

**crosstab:** A report format that compares two intersecting sets of data. For example, you can generate a crosstab report that shows the costs of critical tasks that are running late.

**cumulative cost:** The planned total cost to date on a particular task. This calculation adds the costs already incurred on a task to any planned costs remaining for the uncompleted portion of the task.

**cumulative work:** The planned total work on a particular task. This calculation adds the work completed on a task to any planned work remaining for the uncompleted portion of the task.

**current date line:** The vertical line in a Gantt chart that indicates today's date and time. *See also* Gantt Chart view.

**deadline date:** A date you assign to a task that doesn't constrain the task's timing. However, if a deadline date is assigned, Project displays an indicator symbol if the task runs past the deadline.

**deliverable:** A product or capability necessary to complete the project.

**dependency:** A timing relationship between two tasks in a project. A *dependency link* causes a task either to occur before or after another task, or to begin or end at some point during the life of the other task.

**duration:** The total number of work periods it takes to complete a task.

**duration variance:** The difference between the planned (baseline) task duration and the current estimated task duration, based on activity to date and any remaining activity still to be performed.

**EAC (estimate at completion):** The total planned cost for a specific task, work package, or project. This calculation combines the costs incurred to date with costs estimated for a task's remaining work.

**earned value:** A reference to the value of work completed. A task with $1,000 of associated costs has an earned value of $750 when it's 75 percent complete. *See also* BCWP.

**earned value management:** A project management technique that compares scope, schedule, and cost to measure project performance.

**effort:** The number of labor units required to complete a task.

**effort-driven:** A setting for an auto-scheduled task that requires an assigned amount of effort to be completed. When you add resources to an effort-driven task, the assigned effort is distributed among the task resources equally.

**enterprise custom fields:** Custom fields stored in a global file. These fields can be used to standardize Project content across an organization.

**exception:** A specified date or date range that isn't governed by the default working time calendar.

**expand:** To open a project outline to reveal both summary tasks and subtasks.

**expected duration:** An estimate of the actual duration of a task, based on work performance to date.

**external task:** A task in another project. You can set links between tasks in your project and external tasks.

**File tab:** The tab at the far left end of the Ribbon that you can click to access file- and program-related actions, such as opening a file or setting Project options, in the Backstage view. *See also* Backstage view.

**finish date:** The date on which a project or task is estimated to be — or actually is — completed.

**finish-to-finish relationship:** A dependency relationship in which the finish of one task determines the finish of another task.

**finish-to-start relationship:** A dependency relationship in which the finish of one task determines the start of another task.

**fixed duration:** For auto-scheduled tasks, a task type setting that specifies that the length of time required to complete a task remains constant no matter how many resources are assigned to the task. A half-day seminar is an example of a fixed-duration task.

**fixed-unit:** A task type setting for auto-scheduled tasks that specifies that the resource units are constant; if you change the duration of the task, resource units don't change. This is the default task type.

**fixed-work:** A task type setting for auto-scheduled tasks for which the number of resource hours assigned to the task determines its length. Fixed-work tasks can't be effort-driven.

**float:** The amount of time that you can delay a task before the task becomes critical. Float is used up when any delay in a task will delay the overall project deadline. Also known as *slack.*

**Gantt Chart view:** A standard Project view that displays columns of task information alongside a chart that shows task timing in bar-chart format.

**generic resource:** A type of resource that allows you to make skill-based assignments based on a skill/code profile.

**grouping:** The organization of tasks by a customized field to summarize costs, duration, or other factors.

**ID number:** The number automatically assigned to a task by Project based on its vertical sequence in the project list. The ID number appears at the far left of the task's row.

**inactive task:** A task that still appears in the task list but is struck out and no longer considered in schedule calculations. Marking a task as inactive enables you to document planned activities that ultimately weren't required for the project.

**indent:** To move a task to a lower level of detail in the project's outline hierarchy.

**lag:** The set amount of delay that occurs in a dependency relationship between tasks. In other words, adding lag time creates a delay between tasks.

**leveling:** A calculation used by Project that modifies resource work assignments for the purpose of resolving resource conflicts.

**linking:** To establish a connection between tasks in separate schedules so that task changes in the first schedule are reflected in the second; also, to establish dependencies among project tasks.

**manually scheduled:** The new task mode where Project does not recalculate the task schedule based on changes in dependent tasks. You must manually reschedule tasks when using this mode. This is the default task scheduling mode in new project files.

**material resources:** The supplies or other items used to complete a task (one of two resource categories; the other is work resources).

**milestone:** A task that usually has zero duration, which marks a significant event in a schedule. By default, the Gantt chart shows a diamond marker for a milestone task.

**mini-toolbar:** A toolbar that appears along with a shortcut menu to offer options such as formatting choices.

**network diagram:** A visual display of how all the tasks relate to one another.

**Network Diagram view:** An illustration that graphically represents workflow among a project's tasks; one of Microsoft Project's standard views.

**node:** In Network Diagram view, a box that contains information about individual project tasks.

**nonworking time:** The time when a resource isn't available to be assigned to work on any task in a project.

**outdent:** To move a task to a higher level in a project's outline hierarchy.

**overallocation:** When a resource is assigned to spend more time on a single task or a combination of tasks occurring at the same than that resource's work calendar permits.

**overtime:** Any work scheduled beyond a resource's standard work hours. You can assign a different rate from a resource's regular rate to overtime work.

**parametric estimate:** An estimating method that uses a mathematical model to determine task or project duration.

**percent complete:** The amount of work on a task that has already been accomplished, expressed as a percentage.

**PERT (Program Evaluation and Review Technique) chart:** A standard project management chart that indicates workflow among project tasks. This is a *network diagram* in Project. *See also* Network Diagram view.

**predecessor:** In a dependency link, the task designated to occur before another task. *See also* dependency *and* successor.

**priority:** A ranking of importance assigned to tasks. When you use resource leveling to resolve project conflicts, priority is a factor in the leveling calculation. A higher-priority task is less likely than a lower-priority task to incur a delay during the leveling process. *See also* resource leveling.

**progress lines:** In Gantt Chart view, bars that overlap the baseline taskbar and allow you to compare the baseline plan with a task's tracked progress.

**project:** A unique venture undertaken to produce distinct deliverables, products, or outcomes.

**project calendar:** The calendar assigned to the project file. All new tasks use this calendar by default unless you assign a different calendar to a particular task. The project calendar can be one of the default calendars built into Project — Standard, 24 Hours, or Night Shift — or a custom calendar that you create.

**project management:** The practice of organizing, managing, and controlling project variables to meet the project outcomes and mission.

**Project Server:** A network-based tool for managing projects across the enterprise. A project manager can use Project Professional to publish a project to Project Server so team members and stakeholders can then view and work with the project data.

**Quick Access toolbar:** A small, customizable toolbar near the Ribbon that offers buttons for the most frequently used commands.

**recurring task:** A task that will occur several times during the life of a project. Regular project team meetings or quarterly inspections are examples of recurring tasks.

**resource:** A cost associated with a task. A resource can be a person, a piece of equipment, materials, or a fee.

**resource-driven:** A task whose timing is determined by the number of resources assigned to it.

**resource leveling:** A process used to modify resource assignments to resolve resource conflicts.

**resource pool:** A group of resources created in a centralized location that multiple project managers can access and assign to their projects.

**resource sharing:** A feature that allows you to use resources you created in a resource pool in your current plan.

**retrospective:** A meeting that occurs at the end of each sprint where the team discusses their work and results and looks for ways to improve outcomes. *See also* sprint.

**Ribbon:** The bar at the top of Project that organizes commands on tabs.

**risk:** An uncertain event or condition that, if it occurs, affects the schedule (or another project objective, such as cost, resources, or performance).

**roll up:** A summary-level task that contains all the subtask values.

**rolling wave planning:** Progressively elaborating the amount of detail in the plan for near-term work and keeping farther-out work at a higher level.

**schedule variance (SV):** The difference between the planned and actual duration or the planned and actual finish dates. When using earned value management, SV is the difference between the earned value and the planned value, shown as SV = EV–PV.

**scope:** The work needed to produce the deliverables, products, or outcomes for the project.

**Scrum master:** A servant leader who focuses on supporting the project team.

**shortcut menu:** A contextual menu that presents applicable commands when you right-click an item in Project.

**slack:** The amount of time that you can delay a task before the task becomes critical. Slack is used up when any delay in a task will delay the overall project deadline. Also known as *float*.

**split tasks:** Tasks that have one or more breaks in their timing. When you split a task, you stop it partway and then start it again at a later time.

**sprint:** A short, time box period in which to accomplish work.

**stakeholders:** Those people who are affected by or can affect the project.

**start date:** The date on which a project or task begins.

**start-to-finish relationship:** A dependency relationship in which the start of one task determines the finish of another task.

**start-to-start relationship:** A dependency relationship in which the start of one task determines the start of another task.

**subproject:** A copy of a second project inserted into a project. The inserted project becomes a phase of the project into which it is inserted.

**subtask:** A task that details a specific step in a project phase. This detail is rolled up into a higher-level summary task. A subtask is also called a *subordinate task*. *See also* roll up.

**successor:** In a dependency relationship, the task whose schedule is dependent on the linked predecessor task's schedule. *See also* dependency.

**summary task:** In a project outline, a task that has subordinate tasks. A summary task rolls up the details of its subtasks and has no timing of its own. *See also* roll up.

**tab:** A part of the Ribbon that you can click to display related tools and commands.

**task:** Also known as an *activity,* an individual step performed to reach a project's goal.

**taskbar:** A bar that graphically shows a task's schedule according to the Timeline on the chart portion of Gantt Chart view. A taskbar is also called a *Gantt bar*.

**Task Board:** A board that shows the work to be done and the status of the work.

**Team Planner view:** A view that enables you to see and change assignments in a Timeline-like format.

**template:** A format in which a file can be saved. The template saves elements such as calendar settings, formatting, and tasks. New project files can be based on a template to save the time involved in reentering settings.

**three-point estimate:** An estimating method that accounts for uncertainty by averaging the optimistic, pessimistic, and most likely estimates. A three-point estimate may be modified by averaging one optimistic, one pessimistic, and four most likely estimates.

**Timeline view:** A view that appears along with some other views to give a simplified picture of the overall schedule.

**timescale:** The area of Gantt Chart view that displays units of time. When the timescale is placed against those units of time, taskbars graphically represent the timing of tasks. *See also* Gantt Chart view.

**tracking:** Recording the actual progress of work completed and the costs accrued for a project's tasks.

**variable rate:** A shift in resource cost that can be set to occur at specific times during a project. For example, if a resource is expected to receive a raise or if equipment lease rates are scheduled to increase, you can assign variable rates for those resources.

**WBS (work breakdown structure):** A hierarchical representation of the project work. Each level represents a lower level of detail.

**work resources:** The people or equipment that performs work necessary to accomplish a task. *See also* material resources.

**workload:** The amount of work that any resource is performing at any given time, taking into account all tasks to which the resource is assigned.

# Index

## A



## B

## C

# D

## E

# F

## G

## H



## I



## K

## L

## M



## N

## O



## P

# Q



# R

# S

# T

## U



## V

## W



## Z

## About the Author

**Cynthia Snyder Dionisio** is a well-known project management speaker, consultant, and trainer. She was the project manager of the team that updated PMI's *Project Management Body of Knowledge*, Fourth, Sixth, and Seventh Editions. She is the author of many books, including *A Project Manager's Book of Forms: A Companion to the PMBOK Guide*, and *A Project Manager's Book of Tools and Techniques* (all from Wiley). Her books have been translated into several languages. Cynthia provides consulting services focusing on project management maturity, and she is a much-sought-after trainer in the private, public, and educational sectors.

## Author's Acknowledgments

Publishing a book is not a solo endeavor. It takes many people to transfer information from the author's head into the reader's hands. If you're reading this, it is because the following people at *For Dummies* and Wiley have been instrumental in getting this information to you.

I am grateful to my acquisitions editor, Steve Hayes, for giving me the opportunity to write this book. Kezia Endsley is an excellent project editor — I appreciate her time and effort associated with keeping the schedule on track and helping me with the artwork. My technical editor, Guy Hart-Davis, was exceptional — I appreciate all his suggestions that made this a better book.

Finally, I would like to acknowledge all the students I have had the opportunity to teach. Project management is a wonderful profession, but it is not for the faint of heart. Thank you for allowing me to light the way for you.

## Publisher's Acknowledgments

**Executive Editor:** Steve Hayes
**Managing Editor:** Michelle Hacker
**Project Editor:** Kezia Endsley
**Copy Editor:** Kezia Endsley
**Technical Editor:** Guy Hart-Davis
**Proofreader:** Debbye Butler
**Production Editor:** Mohammed Zafar Ali
**Cover Image:** ©NicoElNino /Shutterstock

# Leverage the power

***Dummies*** is the global leader in the reference category and one of the most trusted and highly regarded brands in the world. No longer just focused on books, customers now have access to the dummies content they need in the format they want. Together we'll craft a solution that engages your customers, stands out from the competition, and helps you meet your goals.

## Advertising & Sponsorships

Connect with an engaged audience on a powerful multimedia site, and position your message alongside expert how-to content. Dummies.com is a one-stop shop for free, online information and know-how curated by a team of experts.

- Targeted ads
- Video
- Email Marketing
- Microsites
- Sweepstakes sponsorship

**20 MILLION** PAGE VIEWS **EVERY SINGLE MONTH**

**15** MILLION **UNIQUE** **VISITORS PER MONTH**

**43%** OF ALL VISITORS ACCESS THE SITE **VIA THEIR MOBILE DEVICES**

**700,000** NEWSLETTER **SUBSCRIPTIONS** **TO THE INBOXES OF** *300,000* UNIQUE **INDIVIDUALS EVERY WEEK**

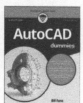

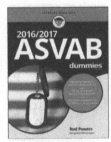

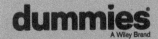